100XN ARCHITECTURAL SHAPE AND SKIN II

HKASP | 先锋空间 主编

100XN 建筑造型与表皮 II

上册

江苏科学技术出版社

PREFACE 序言

Architecture is a kind of spatial art, which uses all kinds of methods rich architectural modeling in limited space, and endows the building with unique vision and feeling, to show the artistic charm. In broad terms, architectural modeling includes functional, economic, technical and aesthetic perfection; As far as the narrow, architectural modeling is a form to construct the building external beauty.

From the perspective of artistic creation, the diversity of the art creation road rich modern architectural modeling form; The progress of science and technology, emergings of new material, use of computer digital construction technology. All of these make the architects' imagination and creativity achieved unprecedented sublimation, and greatly promoted the development of contemporary architectural modeling design.

From the aesthetic point of view, regardless of the ancient and modern, the recognized top grade building must be technical outstanding works. Construction technology and art is as inseparable unified whole, which utilized structure technology to realize the art performance of building modeling clearly, and got the best explanation through the two buildings: Arab Marina + Beach Towers and New York Williamsburg Hotel, the special modeling formed by the exposed structures after artistic treatment have a strong visual impact and infection.

The most difficult problem in Architectural modeling is modeling technique, that is, the overall construction design and part or detail processing form techniques, while the most important way in the part or detail processing is the skin treatment. The relationship between skin and modeling is an eternal topic in the architectural design. Under the impact of new age philosophy and technology, a large number of different appearance and function of the skin, such as the utilize of ecological skin, digital skin and so on make the architectural modeling more colorful; In turn, all kinds of style refine the skin design, strengthen the impact that skin design do for architectural modeling.

The typical significance of modern architectural innovations is that architects can be out of the shackles of traditional architectural form, so that designers can boldly create new buildings adapted to industrialized societies, and make the building shape both in rationalism and radicalism color. The development of society and the progress of science and technology got unprecedented centralized reflect through the contemporary architectural modeling and skin. From Rock museum in Denmark to Singapore National Heart Centre, from CrystalClear Towers Oslo to France Rossignol Global Headquarters, These buildings with different shapes and functions but have a common characteristic: The design link the modeling and skin subtly, and keep delicate balance between the building and environment. They symbolize the qualitative leap on the architecture in the new century.

Building modeling and construction epidermis are closely related, the relationship between them is an eternal topic in the building field, is never-ending search. On that account, we selected 100 cases from the latest new buildings around the world carefully and Edited in book form, hope to provide some reference and appreciates for designers, builders and readers who loved building so much. If you can get a point revelation, that is our great joy.

Editorial Board

建筑是一种空间艺术。设计师运用各种手法在有限的空间里丰富建筑造型,赋予建筑独有的视觉效果和感观体验,展现艺术魅力。就广义而言,建筑造型包括对功能、成本、技术和美学等内容的完善;就狭义而言,建筑造型就是建筑外部形态美的表现形式。

从艺术创作角度看,多元化发展的艺术创作之路丰富了现代建筑造型形式;科技的进步,新材料的涌现,以及计算机数字建筑技术的运用,使建筑师们的想象力和创造力得到了空前的升华,极大地促进了当代建筑造型设计的发展。

从美学角度看,无论古今,公认的顶级建筑都必须是技术上的优秀作品。建筑的技术与艺术是不可分离的统一整体,它要求设计师通过巧妙运用结构技术来实现建筑造型的艺术表现。这种观点在阿联酋"沙湾+沙滩大厦"和纽约"威廉斯堡酒店"这两座建筑上得到了最好的诠释,经过艺术塑造和艺术处理所形成的特殊造型对视觉产生了强烈的艺术冲击力和感染力。

建筑造型中最棘手的问题是造型技巧问题,即总体建筑构思和部分或细部处理形式的技法,而部分或细部处理最重要的莫过于对表皮的处理,表皮与造型的关系是建筑设计中的永恒话题。在新时代理念和新技术的冲击下,大量外观不同和功能各异的表皮,如生态表皮和数字表皮等的运用使建筑造型更加绚丽多彩;反过来,各类风格迥异的建筑造型又更细化了表皮的设计,强化了表皮设计对建筑造型的影响。

现代建筑革新的典型意义在于,建筑师可以摆脱传统建筑形式的束缚,大胆创造出能适应工业化社会的崭新建筑,使建筑造型同时具备了理性主义和激进主义色彩。社会的发展和科技的进步在当代建筑造型和表皮上得到了空前的集中反映。从丹麦"摇滚博物馆"到新加坡"国家心脏中心",从挪威的"透明水晶"雕塑建筑综合体到法国的"ROSSIGNOL 国际总部",这些造型各异、功能多样的新建筑都有着共同特点:巧妙地联系起造型与表皮的关系,并在建筑与环境之间保持微妙的平衡。这些建筑个案象征着建筑造型与表皮在新世纪建筑学上质的飞跃。

建筑造型与建筑表皮息息相关,两者之间的关系是建筑领域里的永恒话题,是永无止境的探寻。有鉴于此,本书从最新涌出的全球众多新型建筑中精心挑选了 100 例编辑成册,希望能为设计师、建设者和热爱建筑的读者提供一些能予以借鉴和欣赏的案例。

编委会

CONTENTS 目录

OFFICE 写字楼

014 Al Nasser Group Corporate Headquarters
Al Nasser 集团总部办公大楼

020 Boulevard Plaza
Boulevard Plaza 大厦

028 M3A2
M3A2 大楼

036 Accra Twin Towers
阿克拉双子塔

042 Office Tower Warsaw
波兰华沙办公大楼

048 SK Networks Daechi-Dong Office Bldg
大峙洞 SK Networks 办公大楼

HEADQUARTER 总部办公楼

058 Bouygues Real Estate Headquarters
布伊格地产集团总部

064 Vidre Negre
意大利 Vidre Negre 办公大楼

072 Iguzzini Ibérica SA Headquarters
Iguzzini Ibérica SA 总部

080 Rossignol Global Headquarters
Rossignol 全球总部

088 Alcatel Head Office
阿尔卡特总部

3M Italy Headquarter 3M 意大利总部	094
Science Park Mechatronik 科技园机电大楼	102
U15 Office Building U15 办公楼	110
Office Building in Pujades 西班牙 Pujades 办公楼	118
Bologna Civia Offices 博洛尼亚市政办公楼	124

市政 CIVICISM

Noain City Hall Noain 市政厅	134
Hasselt Court of Justice Hasselt Belgium 比利时哈瑟尔特市法院	144
Rhone-alpes County Council Hall 法国罗纳 - 阿尔卑斯大区市政厅	150
Sun Moon Lake Administration Office of Tourism Bureau 日月潭风景管理处	158
Congress Center in Krakow 克拉科夫会议中心	164
Prosecutor's Office Tbilisi 第比利斯检察官办公厅	172

180 Embassy of the Republic of Korea in Mongolia
韩国驻蒙古领事馆

192 Tianjin West Railway Station
天津西站

COMMERCIAL PLAZA 商业广场

200 Cet Budapest
布达佩斯 CET 建筑

206 Hanjie Wanda Square in Wuhan
武汉汉街万达广场

214 J Cube
裕冰坊溜冰场

220 Medlacite Liege Belglum
比利时梅迪思购物中心

226 Thaihot City Plaza Mall
泰禾城市广场

234 Importanne Center Sarajevo
萨拉热窝 Importanne 购物中心

COMMERCE AND OTHERS 商业展示及其它

244 Roca London Gallery
乐家伦敦展廊

250 The Culture Yard
文化庭院

260 Automotive Centre of Excellence
卓越汽车中心

266 Vitrahaus
维特拉展馆

Sherbrooke Exhibition Center 274
舍布鲁克展览中心

Genova Traid Fair - Pavilion B 280
热那亚配料展览会 B 展馆

Ferrari World Abu Dhabi 288
阿布扎比酋长国法拉利主题公园

Triose 292
Triose 商业综合体建筑

Markthof Hoofddorp 298
Markthof 综合大楼

酒店 HOTEL

Sofitel Vienna Stephansdom 308
索菲特维也纳斯蒂芬斯顿酒店

The Yas Hotel 318
Yas 酒店

Omni Dallas Hotel 326
欧姆尼达拉斯酒店

Dream Downtown Hotel 334
梦想中心酒店

Torres Porta Fira 344
托雷斯费拉酒店

Chrome Hotel 352
印度铬金酒店

Williamsburg Hotel 358
威廉斯堡酒店

Sanya Hai Tang Bay Mangrove Tree Fairmont Hotel 364
三亚海棠湾酒店

Regent Emirates Pearl 370
摄政埃米尔珍珠饭店

SHAPE INDEX 造型索引

PICTOGRAPHIC SHAPE 象形造型

- 020　Boulevard Plaza　Boulevard Plaza 大厦
- 110　U15 Office Building　U15 办公楼
- 080　Rossignol Global Headquarters　Rossignol 全球总部
- 058　Bouygues Real Estate Headquarters　布伊格地产集团总部
- 042　Office Tower Warsaw　波兰华沙办公大楼
- 218　J Cube　裕冰坊溜冰场
- 318　The Yas Hotel　Yas 酒店
- 352　Chrome Hotel　印度铬金酒店
- 144　Hasselt Court of Justice Hasselt Belgium　比利时哈瑟尔特市法院
- 048　SK Networks Daechi-Dong Office Bldg　大峙洞 SK Networks 办公大楼
- 364　Sanya Hai Tang Bay Mangrove Tree Fairmont Hotel　三亚海棠湾酒店
- 288　Ferrari World Abu Dhabi　阿布扎比酋长国法拉利主题公园
- 266　Vitrahaus　维特拉展馆
- 180　Embassy of the Republic of Korea in Mongolia　韩国驻蒙古领事馆
- 192　Tianjin West Railway Station　天津西站
- 334　Dream Downtown Hotel　梦想中心酒店

OPEN SHAPE 开放式造型

- 036　Accra Twin Towers　阿克拉双子塔
- 094　3M Italy Headquarter　3M 意大利总部
- 280　Genova Traid Fair - Pavilion B　热那亚配料展览会 B 展馆
- 326　Omni Dallas Hotel　欧姆尼达拉斯酒店
- 358　Williamsburg Hotel　威廉斯堡酒店
- 298　Markthof Hoofddorp　Markthof 综合大楼
- 308　Sofitel Vienna Stephansdom　索菲特维也纳斯蒂芬斯顿酒店
- 150　Rhone-alpes County Council Hall　法国罗纳 - 阿尔卑斯大区市政厅
- 274　Sherbrooke Exhibition Center　舍布鲁克展览中心
- 172　Prosecutor's Office Tbilisi　第比利斯检察官办公厅
- 118　Office Building in Pujades　西班牙 Pujades 办公楼
- 260　Automotive Centre of Excellence　卓越汽车中心

博洛尼亚市政办公楼　Bologna Civia Offices　126
泰禾城市广场　Thaihot City Plaza Mall　226

数字造型 DIGITAL SHAPE

Al Nasser 集团总部办公大楼　Al Nasser Group Corporate Headquarters　014
M3A2 大楼　M3A2　028
意大利 Vidre Negre 办公大楼　Vidre Negre　064
Noain 市政厅　Noain City Hall　134
克拉科夫会议中心　Congress Center in Krakow　164
Triose 商业综合体建筑　Triose　292
摄政埃米尔珍珠饭店　Regent Emirates Pearl　370
乐家伦敦展廊　Roca London Gallery　244
布达佩斯 CET 建筑　Cet Budapest　200
武汉汉街万达广场　Hanjie Wanda Square in Wuhan　206
文化庭院　The Culture Yard　250
托雷斯费拉酒店　Torres Porta Fira　344
比利时梅迪思购物中心　Medlacite liege Belgium　218

生态造型 ECOLGICAL SHAPE

日月潭风景管理处　Sun Moon Lake Administration Office of Tourism Bureau　158
萨拉热窝 Importanne 购物中心　Importanne Center Sarajevo　234

负造型 NEGTIVE SHAPE

科技园机电大楼　Science Park Mechatronik　102
Iguzzini Ibérica SA 总部　Iguzzini Ibérica SA Headquarters　072
阿尔卡特总部　Alcatel Head Office　088

SKIN INDEX 表皮索引

ENERGY SKIN 节能表皮

- 102 Science Park Mechatronik 科技园机电大楼
- 094 3M Italy Headquarter 3M 意大利总部
- 110 U15 Office Building U15 办公楼
- 080 Rossignol Global Headquarters Rossignol 全球总部
- 014 Al Nasser Group Corporate Headquarters Al Nasser 集团总部办公大楼
- 180 Embassy of the Republic of Korea in Mongolia 韩国驻蒙古领事馆
- 192 Tianjin West Railway Station 天津西站
- 292 Ferrari World Abu Dhabi 阿布扎比酋长国法拉利主题公园
- 020 Boulevard Plaza Boulevard Plaza 大厦
- 172 Prosecutor's Office Tbilisi 第比利斯检察官办公厅
- 260 Automotive Centre of Excellence 卓越汽车中心

DIGITAL SKIN 数字表皮

- 064 Vidre Negre 意大利 Vidre Negre 办公大楼
- 244 Roca London Gallery 乐家伦敦展廊
- 200 Cet Budapest 布达佩斯 CET 建筑
- 226 Thaihot City Plaza Mall 泰禾城市广场

LIGHT SKIN 轻表皮

- 088 Alcatel Head Office 阿尔卡特总部
- 164 Congress Center in Krakow 克拉科夫会议中心
- 214 J Cube 裕冰坊溜冰场
- 280 Genova Traid Fair - Pavilion B 热那亚配料展览会 B 展馆
- 334 Dream Downtown Hotel 梦想中心酒店
- 344 TORRES PORTA FIRA 托雷斯费拉酒店
- 234 Importanne Centor Sarajevo 萨拉热窝 Importanne 购物中心
- 144 Hasselt Court of Justice Hasselt Belgium 比利时哈瑟尔特市法院
- 364 Sanya Hai Tang Bay Mangrove Tree Fairmont Hotel 三亚海棠湾酒店
- 370 Regent Emirates Pearl 摄政埃米尔珍珠饭店
- 308 Sofitel Vienna Stephansdom 索菲特维也纳斯蒂芬斯顿酒店

维特拉展馆	Vitrahaus	266
舍布鲁克展览中心	Sherbrooke Exhibition Center	272
M3A2 大楼	M3A2	028

透明表皮 TRANSPARENT SKIN

布伊格地产集团总部	Bouygues Real Estate Headquarters	058
波兰华沙办公大楼	Office Tower Warsaw	042
Iguzzini Ibérica SA 总部	Iguzzini Ibérica SA Headquarters	072
阿克拉双子塔	Accra Twin Towers	036
大峙洞 SK Networks 办公大楼	SK Networks Daechi-Dong Office Bldg	048
博洛尼亚市政办公楼	Bologna Civia Offices	126
西班牙 Pujades 办公楼	Office Building in Pujades	118
文化庭院	The Culture Yard	250
Markthof 综合大楼	Markthof Hoofddorp	298
欧姆尼达拉斯酒店	Omni Dallas Hotel	326
Yas 酒店	The Yas Hotel	318
威廉斯堡酒店	Williamsburg Hotel	358
比利时梅迪思购物中心	Mediacite Liege Belgium	220

生态表皮 ECOLOGICAL SKIN

| Noain 市政厅 | Noain City Hall | 134 |
| 日月潭风景管理处 | Sun Moon Lake Administration Office of Tourism Bureau | 158 |

重表皮 HEAVY SKIN

印度铬金酒店	Chrome Hotel	352
法国罗纳-阿尔卑斯大区市政厅	Rhone-alpes County Council Hall	150
Triose 商业综合体建筑	Triose	292

多媒体表皮 MULTIMEDIA SKIN

| 武汉汉街万达广场 | Hanjie Wanda Square in Wuhan | 206 |

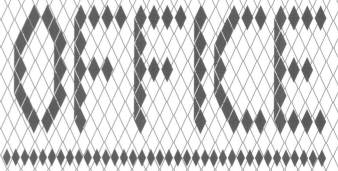

AL NASSER GROUP CORPORATE HEADQUARTERS

Al Nasser 集团总部办公大楼

Architect: ONL (Oosterhuis_Lénárd)
Client: Al Nasser Investments
Location: Abu Dhabi, UAE
Built Area: 21,604 m²
Renderings: ONL (Oosterhuis_Lénárd)

设计公司：ONL (Oosterhuis Lénárd)
客户：Al Nasser 投资公司
地点：阿联酋迪拜
建筑面积：21 604 m²
效果图：ONL (Oosterhuis_Lénárd)

STRUCTURE & MATERIAL 结构与材料

STRUCTURE
Tringulation / Tesselation is used for the structural support system as for the cladding system.

MATERIAL
steel, aluminium, glass

结构
三角形测量 / 曲面细分，用于覆盖系统中的结构支撑系统。

材料
钢材、铝板、玻璃

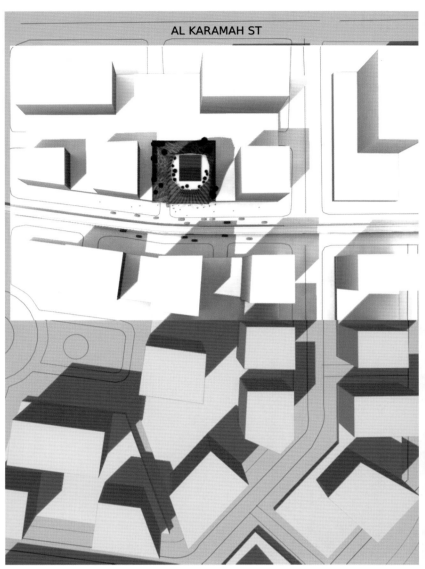

SHAPE ANALYSIS 造型分析

DIGITAL SHAPE

Relatively modest interventions and parametric modifications of the rectangular basic shape while retaining the structural integrity of the design create the iconic appearance of the Al Nasser Group Headquarters Tower. And improves the sustainability in respect to wind loads.

数字造型

本案的建筑设计在保留设计结构完整的同时，对建筑物矩形进行一定的处理和参数修改，使之成为标志性建筑，并提高建筑有关风力载荷的持续性。

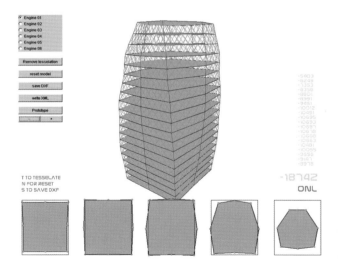

ONL decided to develop a vase shaped tower, narrow at its base, gaining volume in the shaft and tapered towards the top. Since plot C.08 is a plot with a medium sized tower [GF + 1 + 19 Floors] ONL wishes to have a narrower top to allow for more open views from the higher buildings around. The vase has been styled by subtly slicing and chamfering the otherwise rectangular floorplan. Relatively modest interventions and parametric modifications of the rectangular basic shape while retaining the structural integrity of the design create the iconic appearance of the Al Nasser Group Headquarters Tower.

The shape of the tower improves the sustainability in respect to wind loads since the corners of the floorplan, especially in the shaft and the capital part, are chamfered and rounded off.

The window openings count for 20%~25% of the façade surface, at the same time reducing the cooling loss through the windows and reducing the amount of heat coming into the office floors. Silvery transparent coated glass takes care of a good ZTA / LTA value.

ONL 决定设计一座花瓶型的高楼，建筑物基座窄，中部宽，顶部逐渐呈锥形。由于 C.08 是一座总共 21 层的中型高楼，因此，ONL 希望能把建筑物的顶部变窄，使周围较高的建筑能得到更宽广的观景视野。通过断面和斜切等手法，设计师以不规则的矩形平面为建筑打造了独特的花瓶造型。这种对建筑物的矩形进行相对温和的处理和参数修改的方式，在保留设计结构整体性的同时，形成了 Al Nasser 集团总部大楼这种具有标志性的形象。

由于在建筑物平面图的转角处，特别是其脊柱和柱顶部分被去角并修圆磨光，建筑物外形提高了建筑物的风力载荷性能。

立面上的窗户开口面积占表皮的 20%~25%，使建筑能通过窗户较低其内部的制冷损耗，同时减少外部进入办公区的热量。其中，银制的透明玻璃属于 ZTA / LTA 材质。

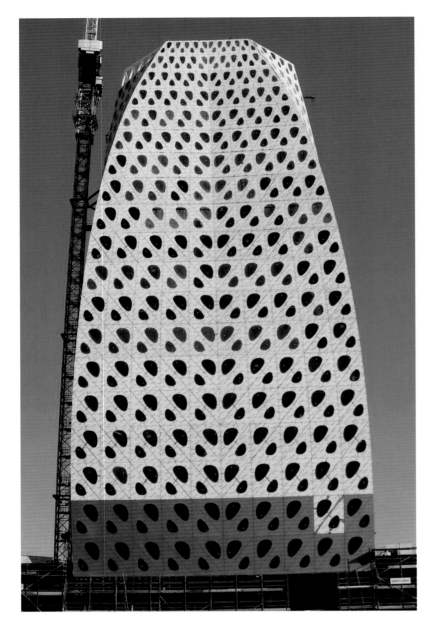

The façade is completely covered with glass, 20%~25% vision glass 75%~80% contrast coloured spandrel panels. The structural glazing technique guarantees minimal concentration of moist dust grains and sets the sound basis for low maintenance costs. Self-cleaning glass will be considered but has to be proofed for the harsh Emirates climate.

All office space are not deeper then 6 m, making all workers profit from direct daylight, and hence saving on lighting costs. The offices are divided in two zones, one closer to the daylight using less artifical light. The zones can be switched on/off separately.

The fully ICT driven file to factory process alows for a maximum control on the produced materials. All component are mass-customized, which means that there will be no waste maetrial whatsoever. Mass-customization offers high-level of sustainability as seen in relation to traditional mass-production methods.

The load-bearing structural façade needs no separate support structure for the façade cladding system. The same tringulation / tesselation is used for the structural support system as for the cladding system. ONL's innovative structural façade system reduces working hours, time and hence money and is naturally susta1inable due to its common sense straightforwardness.

建物立面完全被玻璃覆盖，20%~25% 的可视窗与 75%~80% 的彩色拱肩镶板。结构型玻璃技术有利于减少湿气和粉尘的聚集，为降低维护成本打下了基础。此外，为了适应阿联酋国家严酷的气候环境，设计采用了自动清洁玻璃。

建筑内的办公空间最高不过 6 m，这使得所有工人均可受到日光的直射，从而节省了照明成本。办公室分为两个区域，接近于日光的那个区域将使用较少的人工照明。两个区域的照明单独开关，互不影响。

在建筑材料方面，设计师完全采用了由 ICT 驱动的定制程序，可对成品材料进行最大控制，可以最大限度地定制所有部件，也就是说不会造成材料的浪费。与传统的批量产品生产对比，大批量的定制产品具有较高的可持续性，更为环保。

由于承载结构立面的使用，建筑的骨架外墙不需要独立的支撑结构。另外，该外墙还同样采用了三角形测量 / 曲面细分作为结构的支撑系统。就这样，ONL 的创新结构立面系统减少了建筑时间，节省了建筑成本，还通过其直接、简练的设计手法，体现了建筑的可持续性。

SKIN ANALYSIS 表皮分析

ENERGY SHAPE

20%~25% of the building are windows that can make the most of the areas enjoy the sunshine; the other 75%~80% are color spandrel panels that can effectively reduce quantity of heat to access, reduce the energy consumption of refrigeration.

节能表皮

建筑表皮 20%~25% 的面积为幕窗，能使大部分区域受到阳光直射，另外 75%~80% 则为彩色拱肩镶板，能有效地降低热量的进入，减少制冷能耗。

Boulevard Plaza 大厦

Architect: Aedas
Client: Emaar Properties
Location: Dubai, UAE
Site Area: 17,200 m²

设计公司：凯达环球
客户：Emaar Properties
地点：阿联酋迪拜
占地面积：17 200 m²

STRUCTURE & MATERIAL 结构与材料

STRUCTURE

Reinforced Concrete Frame Structure

MATERIAL

Façade Glazing, Heat Resistant Materials

结构

钢筋混凝土框架结构

材料

釉面玻璃、隔热材料

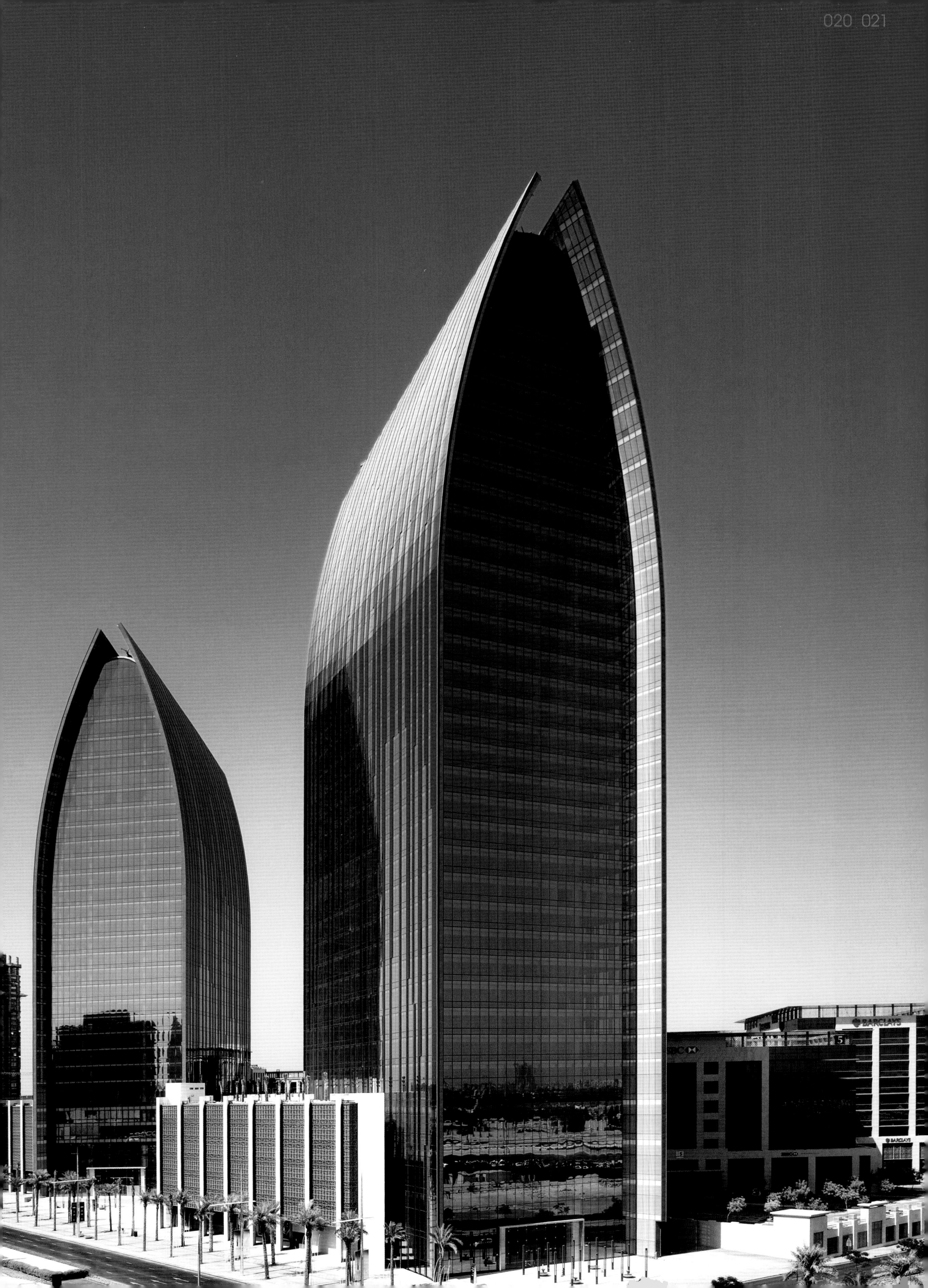
BARCLAYS
BARCLAYS

Boulevard Plaza stands at the gateway into the Burj Dubai development. The importance of the site is accentuated even more by being located immediately across the street from the tallest tower in the world - Burj Dubai Tower.

The design strives to fit appropriately into this development as a respectful icon to the community. The relationship of the forms and their articulation derive from both its contextual response and as a symbol representing modern Islamic architecture set appropriately within the most modern Islamic city in the world – Dubai. Both towers point toward the main entry to greet the visitors. As one continues into the site, the towers rotate their orientation as a gesture of respect to the lofty neighbor across the street. Despite the changing curved form of the building section, the units are modularized to standard layouts for simple construction, rational space and cost efficiency. The two towers of 42 and 34 floors contain grade A+ office space, looking out to take advantage of the views toward and around the Burj Dubai.

The modern façade's Islamic patterns not only offer a contextual symbolism but they also act as a sun screen, which significantly reduces heat loads from the intense Dubai sunlight, thus reducing energy consumption produced by mechanical loads. Over-sailing façades cantilever up to 5 meters offering shade to the East and West elevations which contain a more transparent glass than the North and South patterned façades.

Boulevard Plaza 位于迪拜塔前，由于其处于世界第一高楼——迪拜塔的对面，这里的区域位置日显重要。

建筑的设计以尊重其所在区位的人文和环境为原则，力求能融入其周围环境中。作为世界上最现代化的伊斯兰城市——迪拜的现代伊斯兰地标建筑，建筑各部分的组合模式和接合方法深受当地文化影响。双塔都朝向主入口，似乎在欢迎游客的到来。作为该区域的增建部分，双塔与街道对面的高尚社区两相对望，相互辉映。除了建筑某些部分的弧形设计，建筑内各单位均采取模块化设计，布局标准、结构简单，既能使空间合理化又节约成本效益。分别高 42 层和 34 层的双塔内还包含甲级写字楼，坐拥壮观的迪拜塔及其周边无限美景。

建筑的立面设计极富现代感，上面的伊斯兰图纹样式使建筑成为该地区的标志，同时也充当了日光屏的作用，大幅降低了迪拜灼热阳光下的热负荷，并减少了机械负载的能量消耗。东、西立面采用了比带有图纹样式的南、北立面更为透明的玻璃幕墙，而高出立面 5 m 的悬臂则恰恰为东、西立面提供了遮阴。

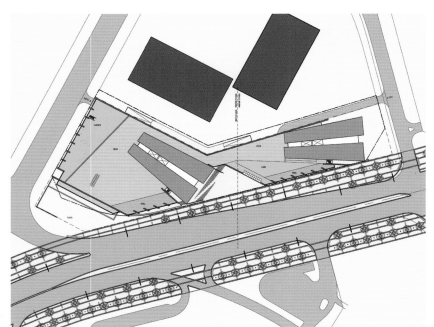

SHAPE ANALYSIS 造型分析

PICTOGRAPHIC SHAPE

The towers are clothed with an articulated skin recalling the veils and layers of traditional Islamic architecture. As the figures rise, they bend inwards, forming two deep, shadowed arches up to the sky and beyond – toward the top of the Burj Dubai.

象形造型

双塔的立面由里外两层表皮构成,让人联想起传统伊斯兰建筑的层式和层次。随着高度的上升,表皮逐渐向内弯曲,形成两个深弯拱门,开口向上直指天际,仿佛正在努力超越远处迪拜塔的顶端。

SKIN ANALYSIS 表皮分析

ENERGY SKIN

Glass façade on full ranges will filter intense radiation on sunny days and introduce adequate lights into interior on cloudy days. Windows can be closed or opened according to temperature changed at any time, so that the building will reduce energy consumption without much reliance on artificial facilities.

节能表皮

全方位的玻璃立面在晴天时能过滤强烈的阳光辐射,阴天时又能将充足的光线引入室内。此外,其窗口还能依据温度变化随时关闭或打开,让建筑无需过多依赖人工设施,从而降低能耗。

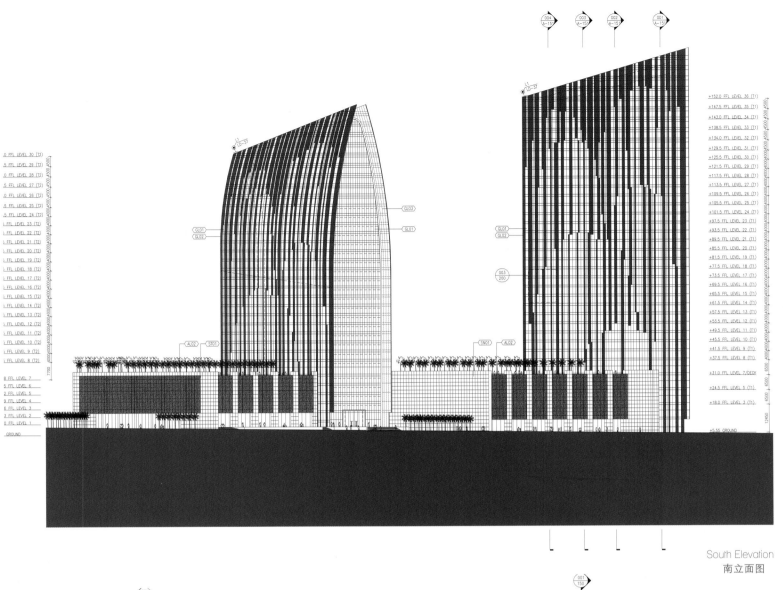

South Elevation
南立面图

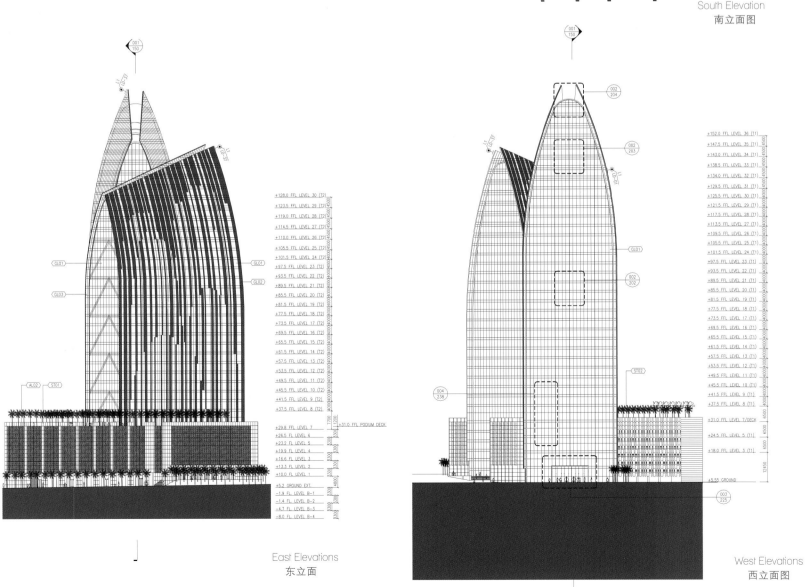

East Elevations
东立面

West Elevations
西立面图

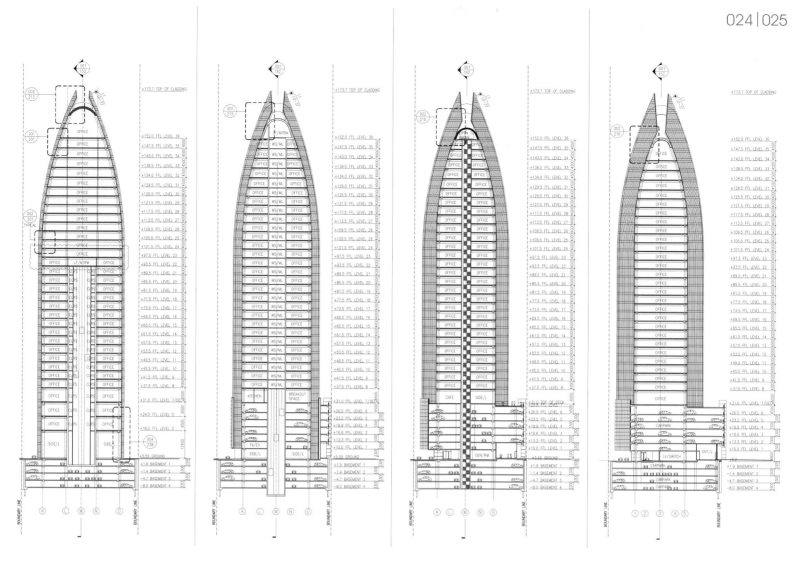

Typical Cross Sections
典型横截面图

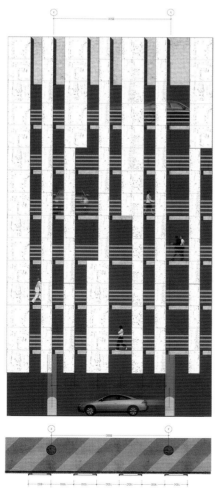

Elevation Details
立面图细节图

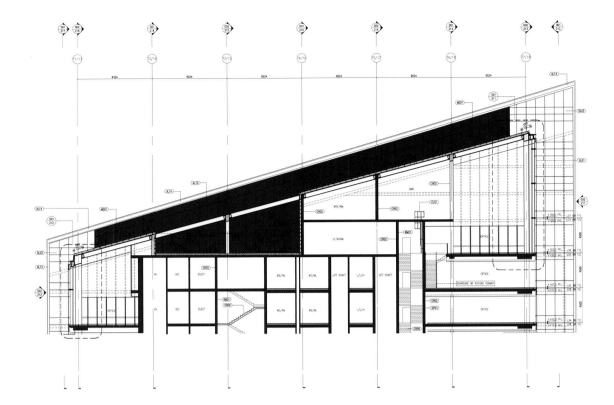

Typical Crown Longitudinal Section
典型顶部纵剖图

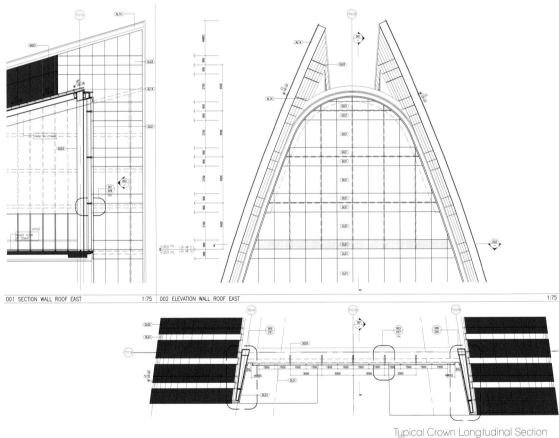

| 001 SECTION WALL ROOF EAST | 002 ELEVATION WALL ROOF EAST | 1:75 |

Typical Crown Longitudinal Section
典型顶部纵剖图

The design considers the structural efficiency by pushing shear walls to the outside of the corridor, therefore widening the structural base and reducing the span between the core and the façade. This effectively reduces the structural depth and construction materials required for the structural members, thereby reducing the embedded energy consumed. The larger core allows pressure to be transferred to the foundation over a larger footprint thus reducing quantities of concrete. The Podium is an open air structure with natural ventilation and fans for air flow circulation. What' more, soft landscape on roof surfaces minimizes solar gain and Water features along roof landscaped areas allow cool natural cooling for users.

考虑到建筑的结构效能，设计师将剪力墙设置在走廊的外侧，并以此扩大了结构基地，缩短了结构中心和立面之间的跨度，有效地减少了结构深度和结构部件施工所需的材料，从而降低了嵌入式能源的消耗。承重压力可以借由较大的轴心转移到大型的地基上，从而减少混凝土的使用量。建筑的墩座采用开放式结构，利用自然通风和排气扇促进空气流动和循环。此外，屋顶表面的绿化减少了太阳能的吸收，而水景的设置也为用户带来了一片自然清凉。

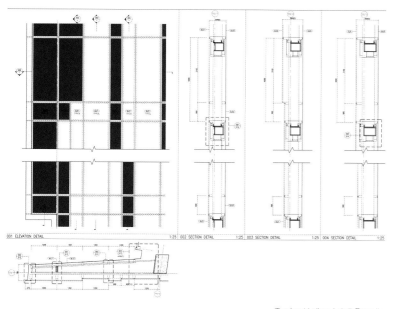

Typical Wing Wall Details
典型翼墙节点图

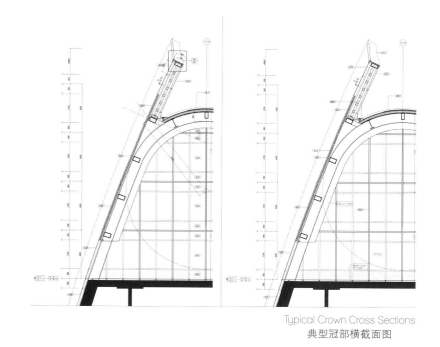

Typical Crown Cross Sections
典型冠部横截面图

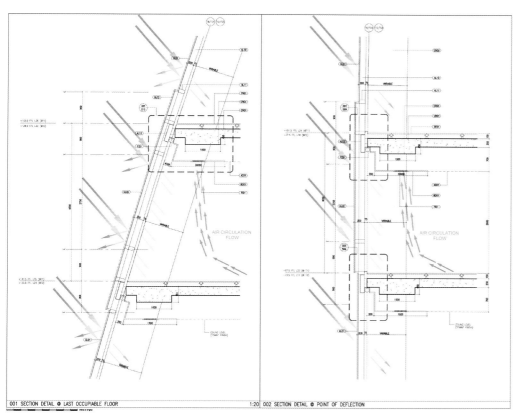

Feature Façade Shading Mechanism Diagram
立面功能遮阳原理图

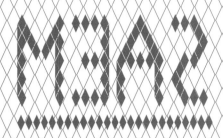

M3A2 大楼

Architect: ANTONINI-DARMON Architects
Location: Paris, France
Site Area: 550 m²

设计公司：ANTONINI-DARMON 建筑事务所
地点：法国巴黎
占地面积：550 m²

STRUCTURE AND MATERIAL 结构与材料

STRUCTURE
Reinforced Concrete Structure

MATERIAL
Steel, Concrete, Glass, Brick

结构
钢筋混凝土结构

材料
钢材、混凝土、玻璃、砖材

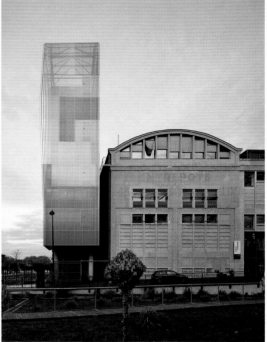

SHAPE ANALYSIS 造型分析

DIGITAL SHAPE

The buildings on the premise of the cultural community of Paris Diderot University fit into the undeveloped, it respects the existing building and accentuates the slenderness of the tower.

数字造型

本案原是巴黎第七大学的文化与社团活动中心,选址在一个未经开发的区域。它在尊重邻近建筑的前提下,以修长的造型引人入胜。

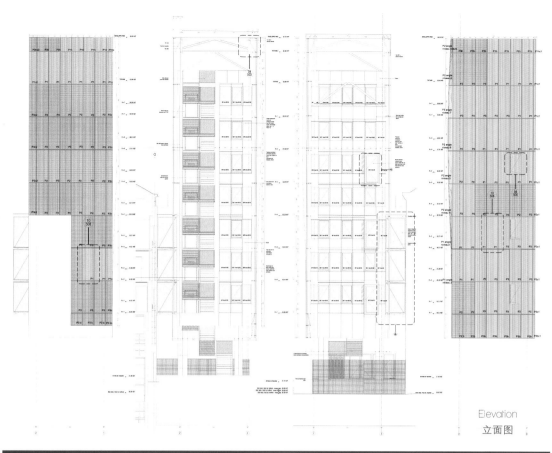

Elevation
立面图

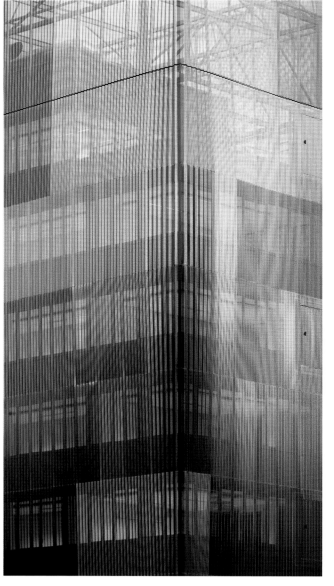

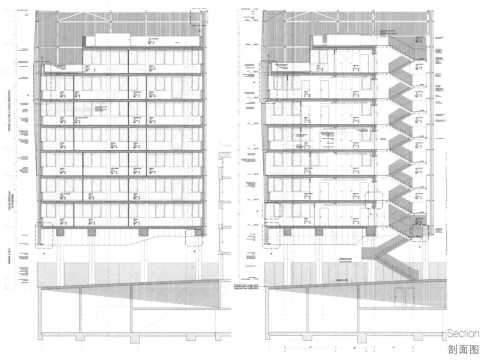

Section
剖面图

SKIN ANALYSIS 表皮分析

LIGHT SKIN

The transparent grid façade is the feature of the building, which is liked a flexible and elegant icon that stands brightly in the neighborhood.

轻表皮

透明的网格立面是本案造型的一大特色,使本案如典雅而轻巧的雕像屹立着,在周边环境中脱颖而出。

The buildings on the premise of the cultural community of Paris Diderot University fit into the undeveloped, southwest area of the Flour Market which is recently converted by Nicolas Michelin and Associates Agency. A break between the Flour Market and the new building is preserved. It respects the existing building and accentuates the slenderness of the tower. The two independent buildings coexist completely.

The signal-like extension stands out of its context by means of its evolving shape. It is a sensitive, delicate object, treated simply to avoid rivalry with the strong presence of the Flour Market. On the contrary, it acts as a light, gravitational counterpoint. An architectural dialectic and emulation come into play much like a castle and its keep, both intrinsically inseparable.

本案位于面粉市场西南角一处未开发的区域，这里原是巴黎第七大学的文化活动场所，近来由 Nicolas Michelin and Associates Agency 建筑事务所进行改造。其实面粉市场与本案的周围留有间隙，本案并没有破坏面粉市场的建筑形态，只是以修长的楼身造型将其与面粉市场区分开来。这种设计方式使得两幢相对独立的大楼完美共存。

本案通过修长的建筑形体在周围环境中脱颖而出，给人以深刻的印象。设计师细腻而灵敏，只经过简单的处理就可避免本案与面粉市场的建筑风格产生冲突，还使得本案亮眼夺目如一盏灯，引人入胜如具备一种引力。其实建筑的结构逻辑与竞争能力就好像城堡与它的外观，从本质上来说是不可分割的。

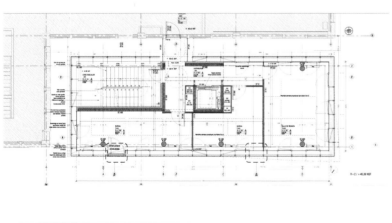
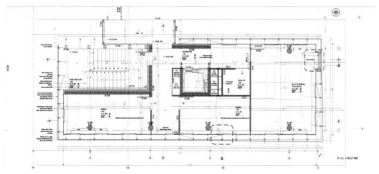
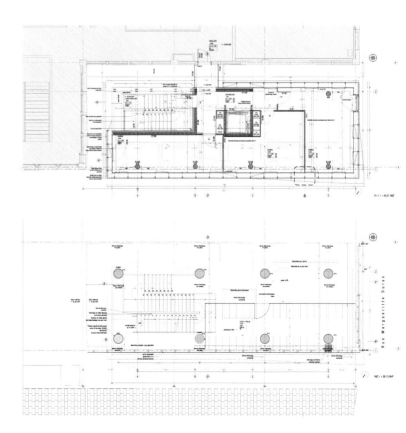

Floor Plan
楼层平面图

ACCRA TWIN TOWERS

阿克拉双子塔

Architect: Frederico Valsassina Arquitectos
Location: Accra, Republic Ghana

设计公司：Frederico Valsassina Arquitectos
地点：加纳共和国阿克拉

STRUCTURE & MATERIAL 结构与材料

STRUCTURE
Reinforced Concrete Structure

MATERIAL
Glass, Monolith

结构
钢筋混凝土结构

材料
玻璃、石料

The present proposal occupies a central position in relation to urban design of the city of Accra. It has place in a regular small sized plot, located on Liberia Road, near to the British Council.

Predicting the program into two towers of 40 floors, the action sought to exploit the small size of the lot in the design process through the theoretical occupation of the entire plot. Corresponding to the maximum volume of occupation, a 160 meters height parallelepiped with a base equal to the construction polygon was conceived as a starting point. Volumes were gradually subtracted from the monolith found, whose absence meant to imply the existence of two autonomous bodies, although formally dependent on each other - the Twin Towers – and to focus on each of the spaces proposed, improving the quality of its essence.

The building is designed to create transition shaded areas between interior / exterior and this concern exists also on a material level. Thus, although the presence of glass is dominant, it exists isolated in the background, in order to protect the interior from sun exposure. In the foreground, in order to emphasize the abstract nature of the intervention, the mesh is introduced as an additional protection, a reflective filter to the outmost intense radiation. The exception occurs in the office tower, breaking the glass element as a surprise factor for the intervention, although the stratification of the plans also protects the interior.

本案是阿克拉城市规划中的中心项目，位于利比里亚大街，靠近英国文化委员会其选址是一块方整的小型场地。

预计整个项目将由两栋 40 层高的塔楼组成，建筑过程主要根据场地的范围采取大量的小规模开发。经过对建筑最大体量的估算，设计师决定构筑一座 160 m 高的平行六面体建筑，其基底与建筑的多边形体相同。建筑的体量会随着整块石料的运用而逐渐减少，暗示着两栋大楼的相对存在感。双子塔尽管在形式上相互依存，却也保持着对各自空间的尊重，以提高建筑的空间质量。

整个建筑的设计以在室内和室外之间打造一个阴影过渡区域为目标，而这在很大程度上取决于材料的应用。因此，尽管同样都是以玻璃作为主材料，但出于对内部空间的保护，防止过于强烈的日照，设计师在建筑背阳面单独使用了玻璃，而为了突出介入面的抽象特征，在建筑的向阳面则增加了网格层，充当阻挡外部强烈辐射的过滤器，起到额外保护的作用。而在办公大楼却并非如此设计，尽管立面的分层也起到了保护建筑内部的作用，却没有特别突出玻璃元素的运用。

SHAPE ANALYSIS 造型分析

OPEN SHAPE

The glass façades set in the shadow of the twin towers help introduce more light and also impose a link between the two towers.

开放式造型

在双塔的背光部位采用玻璃立面，既能为建筑引入更多的光线，同时也强加了两塔之间的联系。

Model 模型图

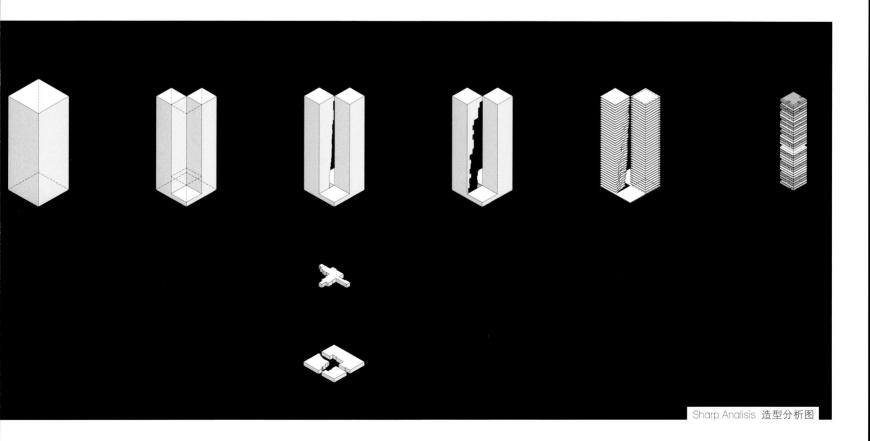

Sharp Analisis 造型分析图

SKIN ANALYSIS 表皮分析

TRANSPARENT SKIN
Glass is dominant and the mesh is introduced as a reflective filter to the outmost intense radiation to be the additional protection.

透明表皮
以玻璃作为主材料，同时采用网格设计，作为过滤器阻挡外部强烈的辐射，起到额外保护的作用。

The coverage of the ground floor as an additional recreational space is restricted to the residential area. Health club, swimming pools, green areas, outdoor terraces and covered seating areas are assumed to be as decompression spaces in order to attenuate the density of the construction and enhance the quality of the solution.

Below this platform, as an extension of street retail, there's a shopping gallery with three different entries, from which offices and housing are acceded. The office tower breaks the ground floor and extends itself into the interior of the gallery, marking its projection on the platform and revealing its presence to visitors; the exterior stretches to reach it and communicates with the void that drives forward the shopping area. The parking lot is projected in six floors below ground level, considering all the assembled functions. The entrance is made in the back of the building, through a common ramp leading to floor -1, serving separately residential and commercial areas.

建筑的首层为一个附加的休闲娱乐场所，仅限于居民区的住户使用。里面配备有健身俱乐部、游泳池、绿地、户外露台和有顶棚的座位区等，是一个理想的放松场所，能起到减少建筑密度并提高生活质量的效果。

在这层建筑平台下方，设置了一条购物街作为街道商店的延伸部分，它带有三个入口，方便人们从办公区和住宅区进入。办公大楼穿过首层，并延伸至走廊内部，在平台上也清晰可见，向游客昭示其存在，其外部则充分利用空间延伸到购物区。考虑到所有的功能分布情况，停车场被设计在底下六层，而入口则被设置在建筑的背面，通过一个斜坡直达负一层，分别服务于住宅区和办公区。

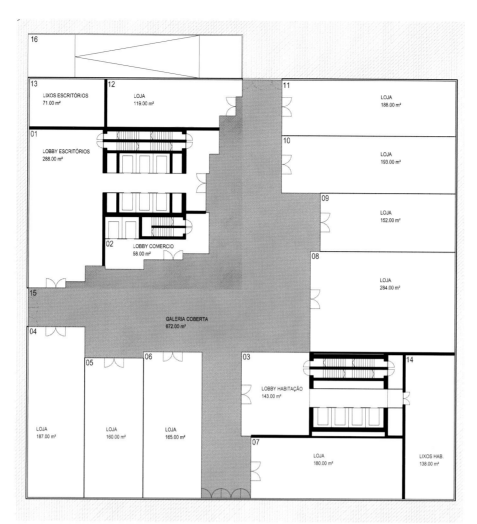

Ground Floor Plan
首层平面图

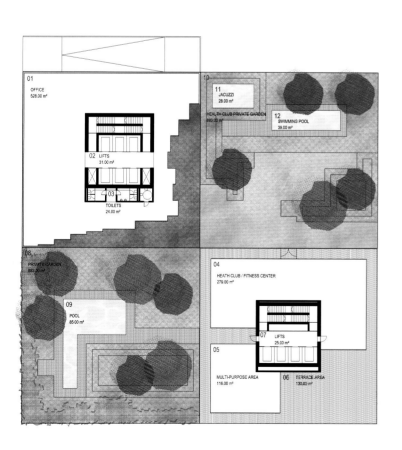

Housing Private Platform
私人住宅平台

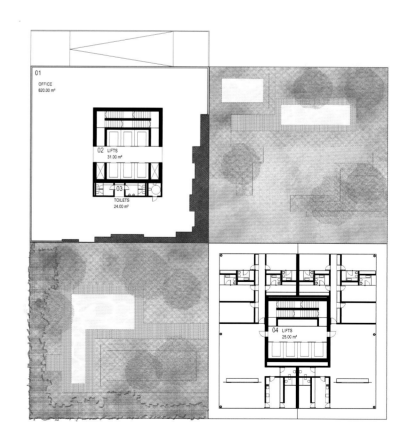

Upper Floor Typical Plan
典型楼层平面图

OFFICE TOWER WARSAW

波兰华沙办公大楼

Architect: Schmidt Hammer Lassen Architects
Client: UBS Real Estate Kapitalanlagegesellschaft mbH
Location: Warsaw, Poland
Site Area: 60,000 m²
Renderings: Schmidt Hammer Lassen Architects

设计公司：Schmidt Hammer Lassen 建筑事务所
客户：UBS Real Estate Kapitalanlagegesellschaft mbH
地点：波兰华沙
占地面积：60 000 m²
效果图：Schmidt Hammer Lassen 建筑事务所

STRUCTURE & MATERIAL 结构与材料

STRUCTURE
Steel Frame Construction

MATERIAL
Steel, Glass

结构
钢架结构

材料
钢材、玻璃

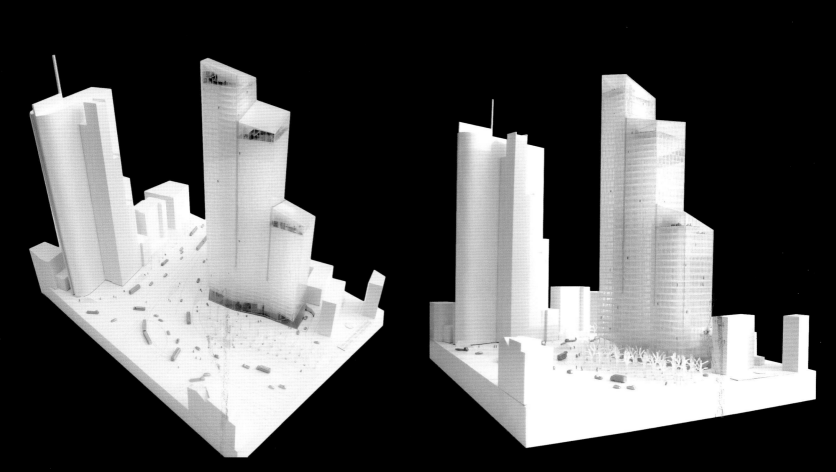

The new Office Tower Warsaw is situated in the financial district of central Warsaw, Poland. The 188 meter high-rise building is to replace the existing 'ilmet' building and will stand out as a modern landmark clearly identifiable in the Warsaw skyline by its unique elegant shape and appearance. The building consists of three individually stepped rectangular volumes with increasing heights towards the east. And the inclined rooflines preserve optimal light conditions for the adjacent buildings.

The design of the building offers a spatial coherence between roof and street level. The lobby at street level, with its spectacular shaped ceiling, corresponds with the sloping shapes of the rooftops, making the building perceive as a sculptural object.

The open lobby allows the people of Warsaw to pass into and through the building, connecting the plaza and park in front of the building with the courtyards of the historical tenement houses to the south. The building offers a number of attractive public areas and serves to complement the project's prominent setting, as well as the entire neighbourhood.

本案由三个独立的矩形体量组合而成。矩形体块形似阶梯，朝着日出之向递增高度，方便建筑物摄取充足的光线。

Model
模型图

SHAPE ANALYSIS 造型分析

PICTOGRAPHIC SHAPE

The building consists of three individually stepped rectangular volumes with increasing heights towards the east, to promise the enough light. While the spectacular shaped ceiling and the sloping shapes of the rooftops, make the building perceived as a sculptural object.

象形造型

本案由三个独立的矩形体量组合而成。矩形体块形似阶梯，朝着日出之向递增高度，方便建筑物摄取充足的光线。广阔奇特的顶棚搭配倾斜式屋顶，给建筑物带来精雕细刻的艺术气息。

SKIN ANALYSIS 表皮分析

TRANSPARENT SKIN

The façade blended in glass elements provides high levels of transparency as well as full integration of sun shading and light reflection shutters. And the sloped rooftops are equipped with photovoltaic cells and elements for rainwater collection, realizing the sustainable development of architecture.

透明表皮

本案在表皮设计上遵循可持续发展的观念。在阻挡日光辐射的设计上，不但采用透明度极高的玻璃作为立面表皮，还辅助使用了太阳能板和百叶窗。此外，倾斜式的屋顶也被安装上光伏电板和雨水集蓄装置。

Ground Floor
首层平面图

The building is designed to reduce energy consumption with the goal of qualifying for the highest levels of sustainability certifications as BREEAM Excellent or LEED Gold status. The modular façade system with floor to ceiling glass elements, provide high levels of transparency as well as full integration of sun shading and light reflection shutters. The sloped rooftops are equipped with photovoltaic cells and elements for harvesting rainwater. The total sustainability approach is a combination of intelligent building management and minimizing technical installations by using passive elements.

本案以降低能源消耗为设计理念，以获得最高等级的可持续认证证书为目标进行设计规划。他们希望得到环境评估守则的肯定或低能电子辐射领域的荣誉证书。外立面所采用的方格状玻璃不但能提高整栋建筑的透明光度，还能与太阳折板和百叶窗一起，起到反射日光的作用。而安装在倾斜式屋顶的光伏电板和雨水集蓄装置，是降低能源消耗的重大举措。此外，各种节能管理方案，智能化装置，也有效地发挥着节能环保的功效。在多种方案的共同协力下，本案才得以成为优秀的可持续发展个案。

SK NETWORKS DAECHI-DONG OFFICE BLDG

大峙洞 SK Networks 办公大楼

Architect: UnSangDong Architects + Junglim Architects
Client: SK Networks
Location: Seoul, Korea
Site Area: 8,267.1 m²
Photography: Bong Kyun Kim

设计公司：韩国韵生同建筑事务所、Junglim 建筑事务所
客户：SK Networks
地点：韩国首尔
占地面积：8 267.1 m²
摄影：Bong Kyun Kim

STRUCTURE & MATERIAL 结构与材料

STRUCTURE
Reinforced Concrete Structure

MATERIAL
Concrete, Steel, Glass

结构
钢筋混凝土结构

材料
混凝土、钢筋、玻璃

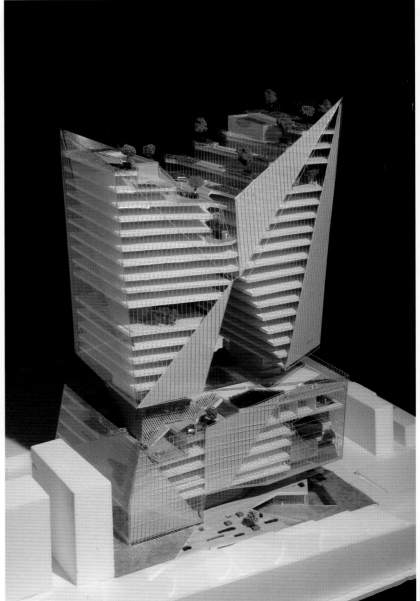

Model
模型图

Program Skin

Energy Skin

Eco Frame

Green Skin

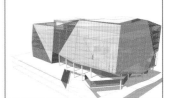
Unique Mass

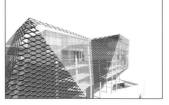
Spectrum Dialogue Facade

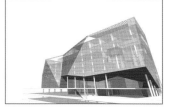
Urban Icon

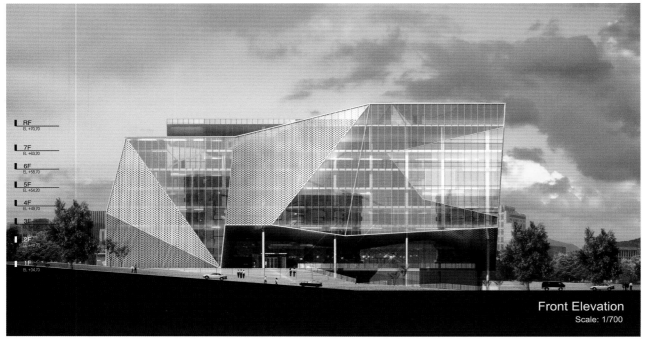

Front Elevation
Scale: 1/700

Elevation
立面图

The inspiration of the SK Networks Daechi-dong Office BLDG is the crystal and diamond. We all know the 4C standard to value the diamond. Cutting makes shapes, create the unique style.The shape of SK is effecting the cutting art.

A floating platform is also new trying, supplying a wide vision and green space for people to enjoy the leisure time. Except the floating art garden outside, there is also an E3 open space in atrium, also an opening sub level, both make the building supply more inter community spaces.

Spectrum dialogue is the design theme. Glaze glass is used to sculpture the crystal and diamond effect, with lots of components indifferent angles, it will shine brightly in the sunshine.

SK 办公室的设计灵感来源于水晶和钻石。众所周知，钻石是以 4 C 标准作为衡量的，其中切割 (Cutting) 可以创造出独特的造型，造就不同价值的钻石。SK 的形状是切割艺术的一种表现。

空中的悬浮式平台是本案一个新的尝试，将提供一个视野开阔的绿色空间供人们欣赏风景，打发休闲时间。除了空中这个悬浮式的艺术花园外，设计师还在中庭设置了一个开放式的电子娱乐空间和一个地下层空间，为建筑提供了更多的活动交流空间。

建筑立面的设计以光谱的应用和影响为主题。釉面玻璃用于雕琢水晶和钻石的闪耀效果，由大量的组件在不同的角度拼凑而成，在阳光下发光时将绚丽夺目。

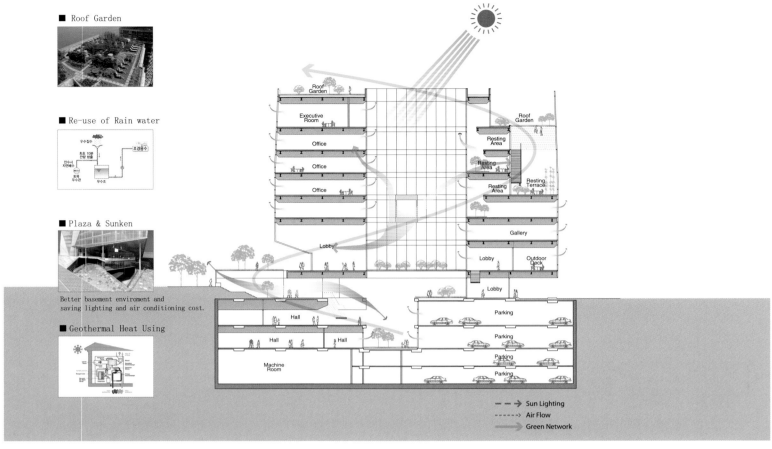

Green Building System
绿色建筑系统

Site Plan
总平面图

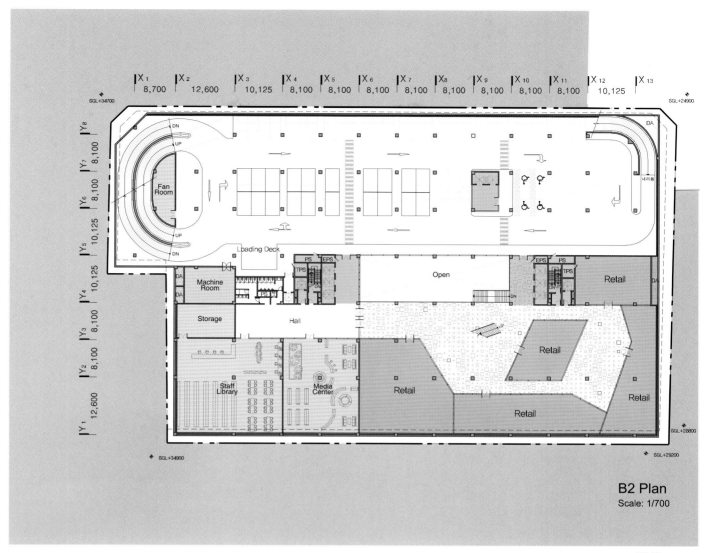

B2 Plan
地下二层平面图

Inter-Community

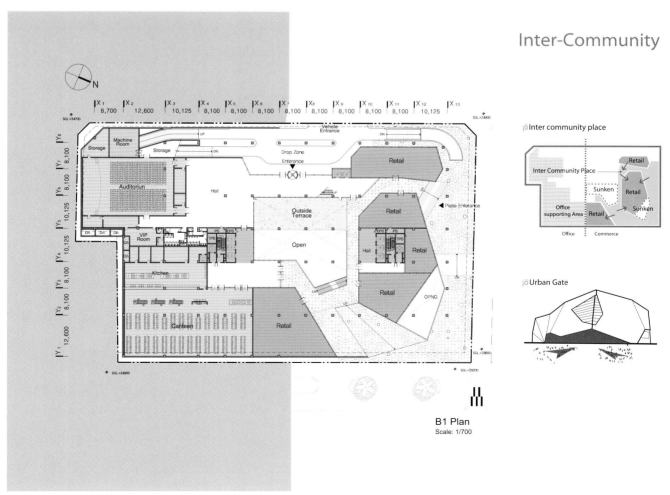

Iner-community Plan
内部社区平面图

SHAPE ANALYSIS 造型分析

PICTOGRAPHIC SHAPE

The shape is inspired by crystal and diamond. Its diamond-shaped façade like the facets of a diamond, shining whether it is day or night.

象形造型

建筑的造型灵感来源于水晶和钻石，菱形的立面如同钻石的切面，不管是在白天还是夜晚，都是那么闪亮。

SKIN ANALYSIS 表皮分析

TRANSPARENT SKIN

Glazed curtain wall endows the building transparency and shiny as crystals, and at the same time meets the lighting needs for the building and plants.

透明表皮

釉质玻璃幕墙的使用，使建筑更具有水晶般的通透和光泽，同时满足了建筑和植物的采光需要。

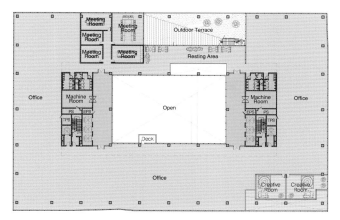

4th Floor Plan
四层平面图

6th Floor Plan
六层平面图

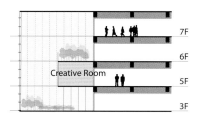

■ Creative room for working people

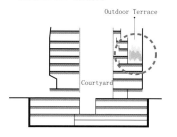

■ Outdoor terrace for the communication between all workers

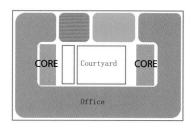

■ Southern East disposition for Pleasant Working Enviroment

Creative Office Zone
创意办公区

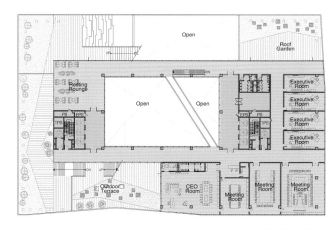

7th Floor Plan
七层平面图

Roof Plan
屋顶平面图

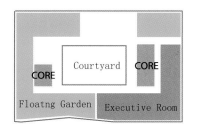

■ Working enviroment with nature

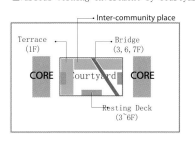

■ Various working enviroment by courtyard

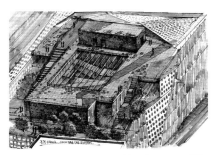

Headquarters Office Zone
总部办公区

HEADQUATER
总部办公楼

BOUYGUES REAL ESTATE HEADQUARTERS

布伊格地产集团总部

Architect: Christian de Portzamparc
Location: France
Built Area: 24,000 m²

设计公司：Christian de Portzamparc
地点：法国
建筑面积：24 000 m²

STRUCTURE & MATERIAL 结构与材料

STRUCTURE
Reinforced Concrete Structure

MATERIAL
Printed Glass

结构
钢筋混凝土结构

材料
喷绘玻璃

The project is composed of three buildings, it's an ensemble which while being urban is formed of two contrasting architectural expressions. The block has been opened, with a public path crossing the site and existing workers lodgings and ancient millstones integrated into a whole.

The Galeo building, Head office for developer in France: Bouygues Immobilier, denotes the entrance of the town of Issy Les Moulineaux from Paris. It constitutes a signal from the quay of the River Seine and the peripherique highway (that encircles Paris), its form and its surface of printed glass scales take on a cinemagraphic dimension. It is a "lamp" visible at night. The entire project is a credited HQE for a high quality environmental performance.

该项目由三座建筑构成，是一座建筑群，其开放式的设计与横穿这片区域的公共小路相得益彰，别有一番风情。

该建筑是法国头号开发商布伊格地产集团的总部，亦是从巴黎进入伊西莱穆利欧这座城镇的地标性建筑，同时充当着塞纳河码头以及环绕巴黎的佩理菲瑞克高速公路的标志，其形象、生动的喷绘玻璃犹如黑夜里一盏可视的明灯。该项目荣获了"HQE 高质量环境认证"的殊荣。

SHAPE ANALYSIS 造型分析

PICTOGRAPHIC SHAPE

The project stands erectly in the open block, and interweave with the public road, all this constitutes a signal from the quay of the River Seine and the Peripherique highway, even the landmark denotes the entrance of issy Les Moulineaux from Paris.

象形造型

本案以建筑群的形式矗立在街区内，呈开放式设计，与公共小路相交织，成为塞纳河码头及佩里菲瑞克高速公路的标志，以及从巴黎进入伊西莱穆利欧的地标性建筑。

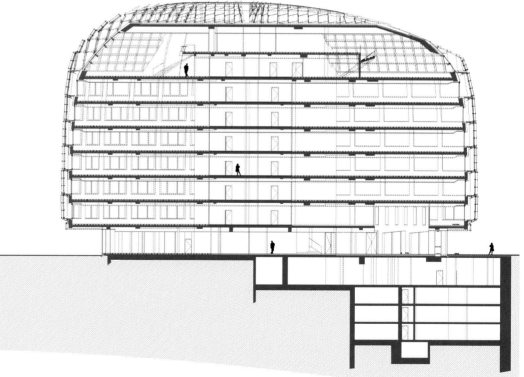

Transversal Section
横剖面图

SKIN ANALYSIS 表皮分析

TRANSPARENT SKIN
The application of printed glass makes the building looked both lively and vividly, as a lamp shining in dark night.

透明表皮
喷绘玻璃表皮的运用使建筑看起来充满活力，在漆黑的夜里闪亮照人。

意大利 Vidre Negre 办公大楼

Architect: Damilano Studio Architects
Client: PORTA ROSSA S.P.A.
Location: Cuneo, Piemonte, Italy
Site Area: 1,400 m²
Photography: Andrea Martiradonna

设计公司：Damilano 建筑事务所
客户：PORTA ROSSA S.P.A.
地点：意大利皮埃蒙特大区库尼奥
占地面积：1,400 m²
摄影：Andrea Martiradonna

STRUCTURE & MATERIAL 结构与材料

STRUCTURE
Concrete Pillars Brick Structure

MATERIAL
Photovoltaic Panels, Oak, Vietnam Black Stone, Glass

结构
砖混支柱结构

材料
光电板、橡木、越南黑色理石、玻璃

Oficina Vidre Negre was born as a contemporary sculpture, in a context of peripheral node of the motorway; the building is set within a park, planted with native species, in which a sort of town square accompanied to the entrance. A prism completely covered in black glass faceted and integrated photovoltaic panels.

本案好比一座当代雕塑,外形呈棱柱体,外立面覆盖着黑色镜面和集成光伏电池板。它坐落在高速公路旁的一个种满原生植物的公园内,其入口与普通的城市广场相若。

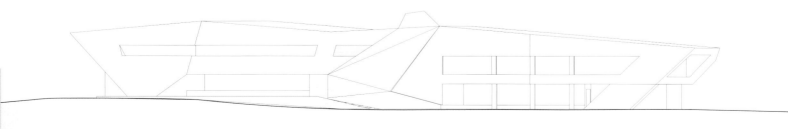

North Elevation
北立面图

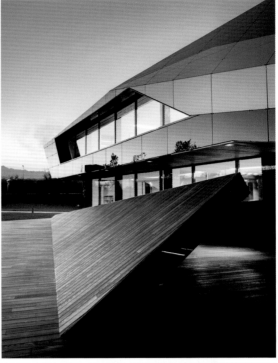

SHAPE ANALYSIS 造型分析

DIGITAL TYPE

The building structure is split, and opens up on the east-west, two wings separated by a central block of facilities.

数字造型

本案采用分裂结构，两个像翅膀一样的体块经过一个中心设备，同时往东西方向向两侧延伸开来。

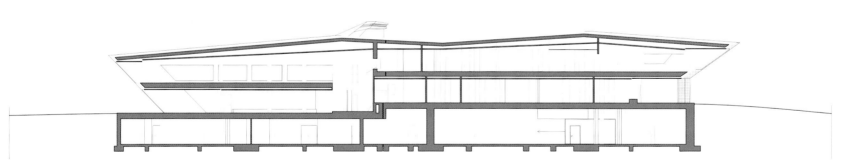

Section A-A
剖面图 A-A

The structure is split reflected in the wide expanse of water over the stage function, is the reserve of water for fire-fighting system and irrigation of the park.

Operating areas are designed as open-space, while the executive offices, the private sector, are tested in the east of the building, projected to the outside and suspended in a vacuum.

本案的分裂结构在功能上主要体现在对大片水域的应用中,为消防灭火和植被灌溉储备水源。

工作区被设计成一个开放的露天空间,而行政办公区和私营部门则设在本案的东侧,像悬浮在空中一般,一直延伸至建筑外部。

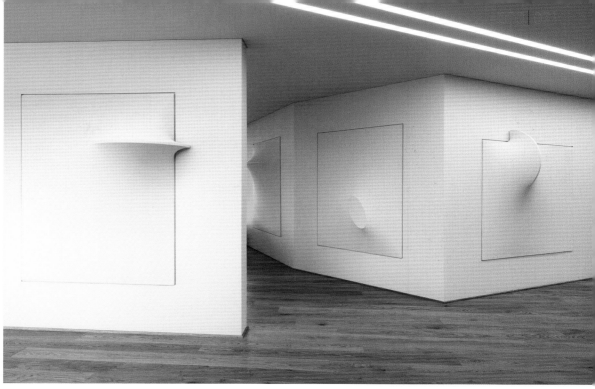

The decomposition of the volume remains in the interior, intersected a sequence of spaces on four staggered levels, illuminated by more than cuts and side windows.

In external floors and ceilings in oak cooked in an autoclave, vietnam black stone floors with large plates of extra-clear glass railings.

本案体量中所采用的解构设计也被用于室内设计中,将建筑中统一的室内空间分割成四个相互交错的房间,通过黑色镜面和侧开的窗户为室内引入自然光线。

此外,高温烫压橡木的露天地板和天花板、黑色的越南大理石地面和透明玻璃围栏,彼此相互映衬。

SKIN ANALYSIS 表皮分析

DIGITAL SKIN

A prism completely covered in black glass faceted and integrated photovoltaic panels. External floors and ceilings made in oak cooked in an autoclave and Vietnam black stone floors with large plates of extra-clear glass railings show a kind of irregular line.

数字表皮

本案外立面被黑色镜面和集成光伏电池板所覆盖,外观展现如棱柱形。由高温烫压橡木打造的露天地板和天花板、黑色越南大理石打造的室内地面,以及透明的玻璃围栏,共筑了一种不规则的线条。

iGuzzini Ibérica SA 总部

Architect: Josep Miàs
Client: iGuzzini Ibérica SA
Location: Barcelona, Spain
Site Area: 9,000 m²
Photography: Adrià Goula

设计公司：Josep Miàs
客户：iGuzzini Ibérica SA
地点：西班牙巴塞罗那
占地面积：9 000 m²
摄影：Adrià Goula

STRUCTURE AND MATERIAL 结构与材料

STRUCTURE
Metal Structure

MATERIAL
Metal, Glass

结构
金属构架

材料
金属、玻璃

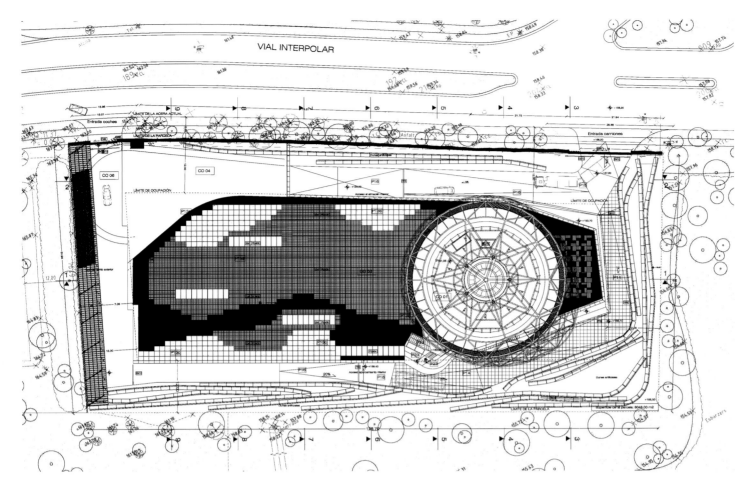

Roof Plan
屋顶平面图

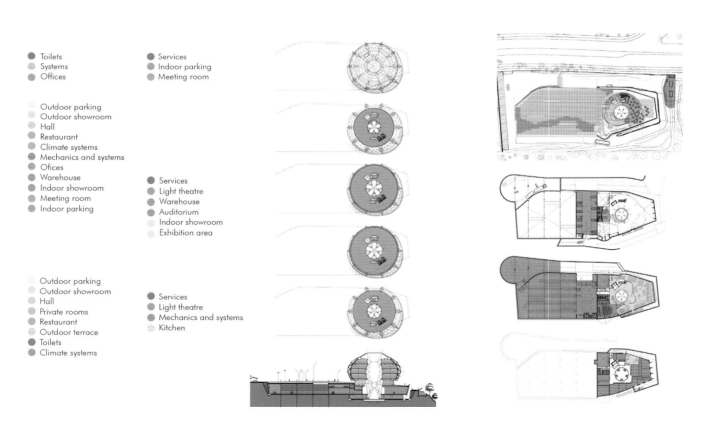

Function Analysis
功能分析图

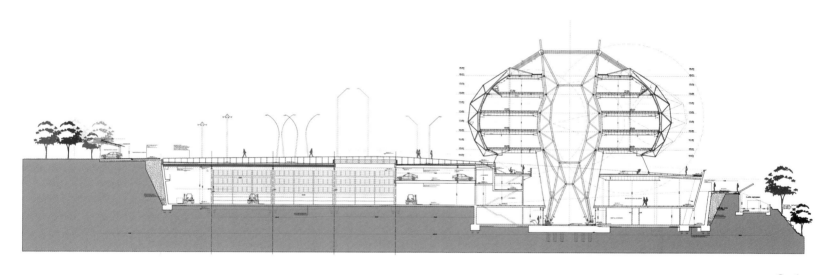

Section
剖面图

The building has two parts, each with a different function: one is low, extensive and underground, with no natural light and built in concrete; the other is spherical and with a glass shell, floating over the landscape.

Taking profit of the slopes of the ground, the platform contains the logistic warehouse, the parking, the showroom, the auditorium, the light theatre, a conference room and climate and system mechanics. All these spaces need darkness, so as to show the characteristics of artificial light.

The surface of the platform is, in fact, a completely equipped outdoor raised floor, covered with different panels which offer the needed flexibility to set up various displays for the outdoor showroom. Over the platform, with an uncertain dynamic equilibrium, stands the more representative area of the complex, a slightly distorted sphere. In this volume, general and management offices are found as well as investigation areas.

建筑物分为两个部分、各具功能、分别建于地下和地上。地下部分体量宽大，无自然采光，地上部分建筑呈球状，罩有玻璃外壳，作为一种景观漂浮于地面上。

建筑的设计充分利用了其坡地地形，将包括物流仓库、停车场、演播室、礼堂、灯光剧场、会议室和调温装置等无需自然光的空间设置在同一个平台中，采用人工照明。

平台表面是一层整装突出的地板，以不同的面板覆盖着，以便室外展示厅能根据需要灵活展示。建筑最突出的部分位于平台上方，是一个稍显扭曲的球体建筑，利用不确定的动态平衡竖立着。在球体空间中设有总管理办公区和调查区。

SHAPE ANALYSIS 造型分析

NEGATIVE SHAPE

The whole building is divided into two parts: above and below the ground. Aerial parts was a sphere shape like a landscape. The other parts that do not need much natural lighting are placed in the ground, so the building will not only protect the environment, but also enhance the quality of the environment.

负造型

整个建筑分为地上和地下两个部分。地上部分呈球体状，具有景观效果；大部分不需要自然采光的空间则被安排到地下，使建筑不但具有了环境保护的意义，更提升了其周边环境的质量。

Actually, the offices are built around a light patio, where the structure is developed: a single pillar formed by five metallic masts. The end of these masts is joined with cables to other ten vertical elements which fix the exterior limit of the slabs. Therefore, the whole building remains hanged only from the central pillar. The offices' façade, covering the external volume and the inner façade of the patio, is a great glass shell. Thus, from the offices, a complete 360° panoramic can be seen. The glass façade is covered over the sunniest surfaces with a solar protector made of a three-dimensional metallic structure, where a special solar fabric is tightened. This textile façade is capable of reflecting the radiation while allowing a great visual permeability. Thanks to its flexible geometry, the solar protection can be very well adapted to the shape and volume of the building.

During the day, from the outside, the building appears like an opaque volume, which reflects the sky on the northern part, and is profiled as a pointed spherical volume on the southern. From the inside, however, the perception is completely different, since the space has no interruption between interior and exterior.

During the night, all these qualities are inverted. Thanks to light, the sphere turns to be almost like a lighthouse, showing its interior in a complete transparency from every corner of the roads which surround the plot.

实际上，球体内的办公区是围着发光的天井而建的，该天井是主结构区，由五根金属柱组成一根主承重柱支撑着。五根金属柱的末端分别使用缆索与十个垂直锚杆相连，锚杆被固定在锚板上，露出一定长度的锚桩。因此，整个建筑可以像"球"一样"挂"在中心柱上。覆盖办公区外立面和内部天井正面的是一个巨大的玻璃壳，从办公室可以看到360°的全景景观。玻璃表面还罩着一层防日光遮光层，由立体金属框构成，上面绷着既透光又防辐射的织物。由于其性能柔韧，可以很好地附着在建筑物表面。

白天，从外面看，建筑物北面反射着阳光，就像个古怪的物体，从南面看，建筑物则似一个带尖的球体。而在建筑物里面，由于空间没有被隔开，感觉完全不同。

到了夜晚，这些特性都被颠倒了。由于灯光的照射，建筑物球体完全像一个灯塔，可从建筑物周围道路的各个角落看到这个完全透明的建筑物的内部。

SKIN ANALYSIS 表皮分析

TRANSPARENT SKIN

The overground part of the building is covered with a layer of glass shell, which can adjust lighting, prevent solar radiation and broaden views. At night, interior LED lights penetrating out the transparent skin endow the building with a changing dynamic.

透明表皮

建筑的地上部分被一层玻璃外壳所覆盖着，既能起到调节光照和防辐射的作用，又能扩宽人们的视野。在夜间，建筑内部亮起的 LED 灯光将穿透外部的透明表皮，赋予了建筑变化多端的动感外观。

ROSSIGNOL GLOBAL HEADQUARTERS

ROSSIGNOL 全球总部

Architect: Hérault Arnod Architectes
Client: Skis Rossignol SAS
Location: France
Site Area: 11,600 m²
Photography: André Morin, Gilles Cabell, Marie Clérin, Christian Rausch, Hérault Arnod

设计公司：Hérault Arnod Architectes
客户：Skis Rossignol SAS
地点：法国
占地面积：11 600 m²
摄影：André Morin、Gilles Cabell、Marie Clérin、Christian Rausch、Hérault Arnod

STRUCTURE AND MATERIAL 结构与材料

STRUCTURE
Steel Structure
MATERIAL
Organic Wood, White Birch, Glass

结构
钢结构
材料
有机木材、白桦树、玻璃

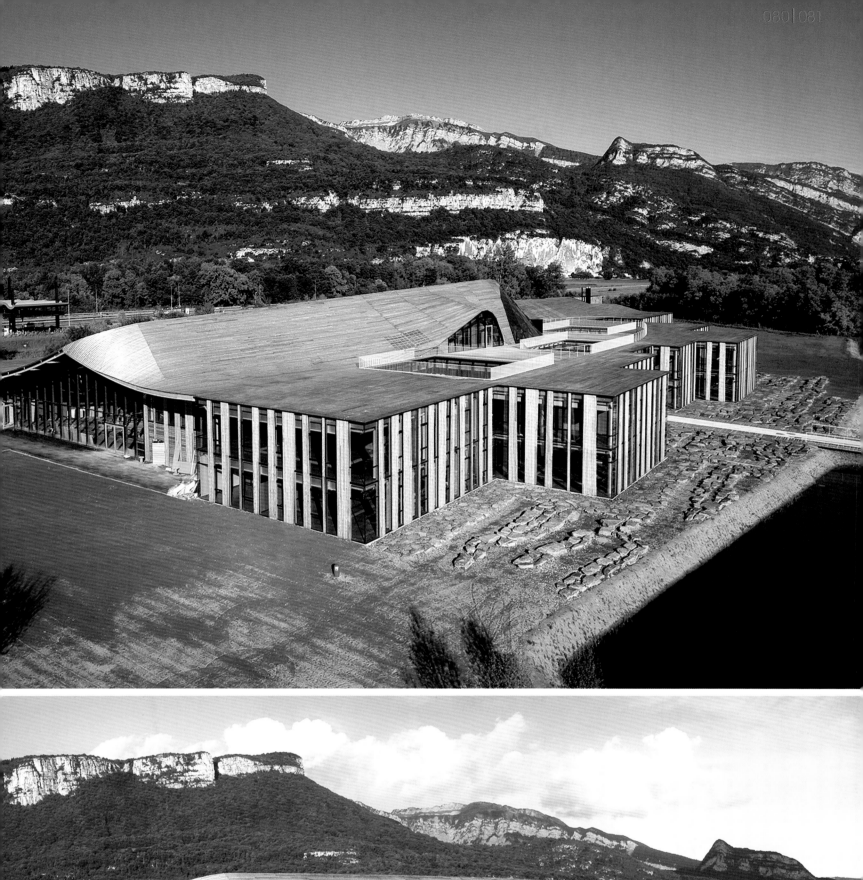

The architecture has been designed specifically for Rossignol, a fusion of the company's functional and fantasy aspects, in a surprising and minimalist form. The roof, which envelops the whole project, is topography in osmosis with nature and the landscape. Its organic, timber-clad shape echoes the profile of the mountains that surround the site.

The front of the building rises to form a roof over the workshops and then on to the apex, and descends again on the south-western side to cover the office area. It is then intercut with patios planted with birch trees that seem to grow through the roof: nature and building intertwine. The irregular profile of the roof and office façades leaves the opportunity for future extensions as required.

本案主要针对 Rossignol 这个品牌进行设计，融合了该品牌的功能和发展愿景，最终打造了一座令人惊喜的简约派风格建筑。屋顶的造型是本案的重点。它依照地理地形，与大自然和外部景观相映成趣。此外，本案使用有机木材为顶棚，所呈现出的形态结构恰似环绕在外的山体轮廓。

本案屋顶造型富有动态美感。屋顶的前部稍微上扬，在盖过工厂车间后，抬升到最高点，随后又往西南方逐渐下沉并覆盖整个办公区域。每一个屋顶轮廓的交界处下方的露台中都种植了白桦树。从整体上看，白桦树好像穿过屋顶向上生长，呈现出大自然与建筑物之间相互交融的和谐画面。此外，不规则的屋顶轮廓和办公大楼的外立面设计为建筑未来的发展留有余地。

SHAPE ANALYSIS 造型分析

PICTOGRAPHIC SHAPE

Rossignol is the pronoun of sliding movement, obviously the building materials and roof modelling design, can let person thought of this brand and culture without consciously.

象形造型

Rossignol 品牌是滑行运动这个领域的龙头老大，可以说它的品牌名字就是滑行运动的代名词。通过本案的材质及其呈现出滑行动态的屋顶造型，人们很容易就能联想到这个品牌和它的文化内涵。

SKIN ANALYSIS 表皮分析

ENERGY SKIN

The external envelope only chooses two materials: organic wood and glass. White birch looked as through the roof, which form a good interaction with the nature. These not only highlights the simplicity idea of the building, but also responded to the building's energy-saving call.

节能表皮

本案外表皮以有机木材和玻璃为主材料，还使用白桦木点缀造型。从远处望去，白桦树看起来像穿过了屋顶一样，与本案的造型在空间形态上形成互动交流。其实这种设计不仅突出了自然简约的建筑风格，还响应了节能建筑的号召。

Only two materials are used for the external envelope: wood (natural larch) and glass. The structure is made of steel, like an organic skeleton that outlines the shape, with its multiple warped surfaces. The workshop space has a primary horizontal roof overlaid by the timber over-roof, creating a hidden space between the two which contains all the technical systems and machinery.

The building is designed for minimal environmental impact. The technical choices make it an efficient and energy-saving building, well insulated and protected from the summer sun by the timber over-roof. The systems are optimised – the heat produced by the workshop machines is recovered and re-injected into the heating network. The offices receive natural ventilation through automatic window opening.

本案外表皮只使用了两种材质——木材（天然落叶松）和玻璃。建筑采用钢筋结构，如同生物骨骼般勾勒出建筑物的外部轮廓，塑造出多种弯曲的形状。工厂车间采用了水平屋顶，其上覆盖着一层带有弧度的木制屋顶。不同形状的屋顶之间形成并巧妙地隐藏了一个空间，而这个空间足以将所有的机械系统和设备都收纳在内。

技术的创新与材质的选择，使建筑本身鲜少受到环境的制约。这无疑是一座使用率极高的环保节能型建筑。首先，屋顶选择了天然的绝缘材料，木材，以保护建筑物免受夏日曝晒。其次，工厂机器运作所产生的热量经过系统优化技术，会被回收并重新注入供热网中使用。最后，办公室装有自动接风窗口，能保证建筑的自然通风。

Section
剖面图

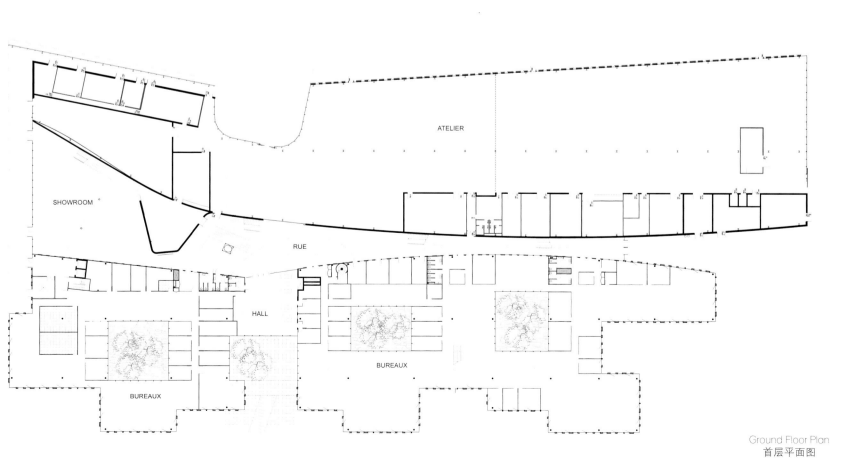

Ground Floor Plan
首层平面图

Inside, two great glass-roofs divide up the panoramicviews to the sky and the mountains. A large roof terrace is available for alfresco lunching, protected from the noise of the motorway.

走进建筑内部，人们可透过两个巨大的玻璃屋将蓝天白云和山脉全景尽收眼底。然后在远离高速公路聒噪吵闹的安静之处——那个大型的顶楼露台上，享用清新舒适的户外午餐。

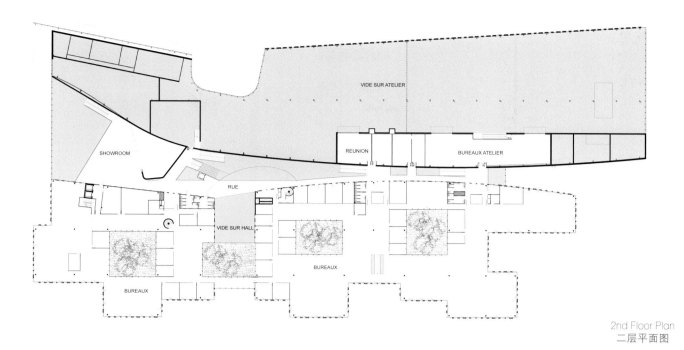

2nd Floor Plan
二层平面图

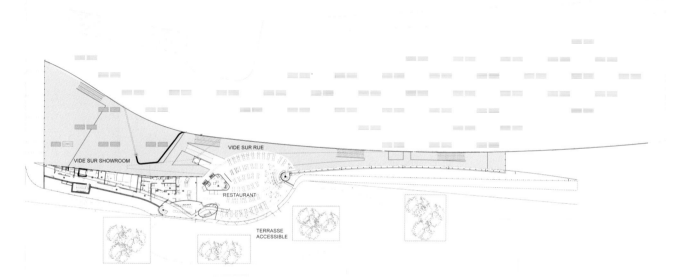

3rd Floor Plan
三层平面图

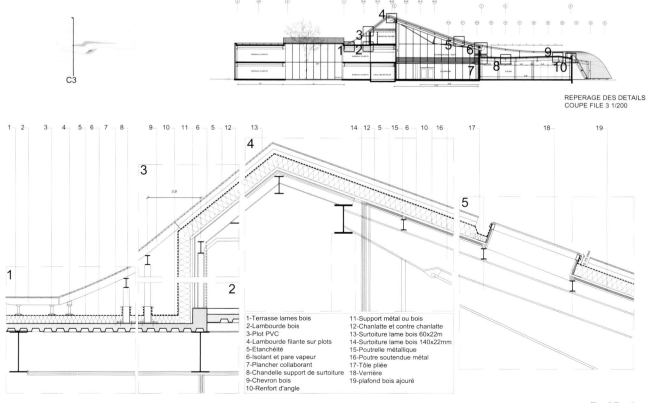

Roof Section
屋顶剖面图

ALCATEL HEAD OFFICE

阿尔卡特总部

Architect: Frederico Valsassina Arquitectos
Client: Garret Properties Inc.
Location: Cascais, Portugal
Built Area: 11,051 m²
Photography: FG+SG - Fotografia de Arquitectura

设计公司：Frederico Valsassina 建筑事务所
客户：Garret Properties Inc.
地点：葡萄牙卡斯卡伊斯
建筑面积：11 051 m²
摄影：FG+SG - Fotografia de Arquitectura

STRUCTURE AND MATERIAL 结构与材料

STRUCTURE
Concrete Columns Structure

MATERIAL
Screen-printed Glass

结构
混凝土梁柱结构

材料
网状印花玻璃

The building appears as an isolated block, morphologically linked to the ones around it, with a distinctively contemporary architectural language.

Using a volumetrically speech similar to the adjoining volumes, the new building is carved in order to soften and project its image to the outside. The form is enhanced as a project-asset, contributing for the intrinsic dynamics that is intended for the global proposal.

The two levels give place to one, through the "folding up" of the entrance level. The new volume releases itself from the ground, as a business card for anyone who enters. Suspended over the void, it directs people to the entrance, which is made to the North over the void that gives access to the parking lot.

整个建筑看上去像是一个独立的体块，在形态上以一种鲜明的当代建筑形象与周边建筑紧密相连。

通过与相邻体块在体积上保持相似，新建筑被切分成块，既柔化了其突兀的形象，又向外界展示了独到的一面。该建筑的外形作为一个项目亮点，对企业内部的全球化发展动态具有一定贡献。

通过入口楼层的"折叠"，设计师采用双层设计取代单层。对于任何进入到这里的人来说，矗立在地面上的新建筑就像是一张名片般，象征着企业的形象。建筑悬浮在空中，仿佛一枚指针，指引着人们进入位于建筑北面、通向停车场的入口。

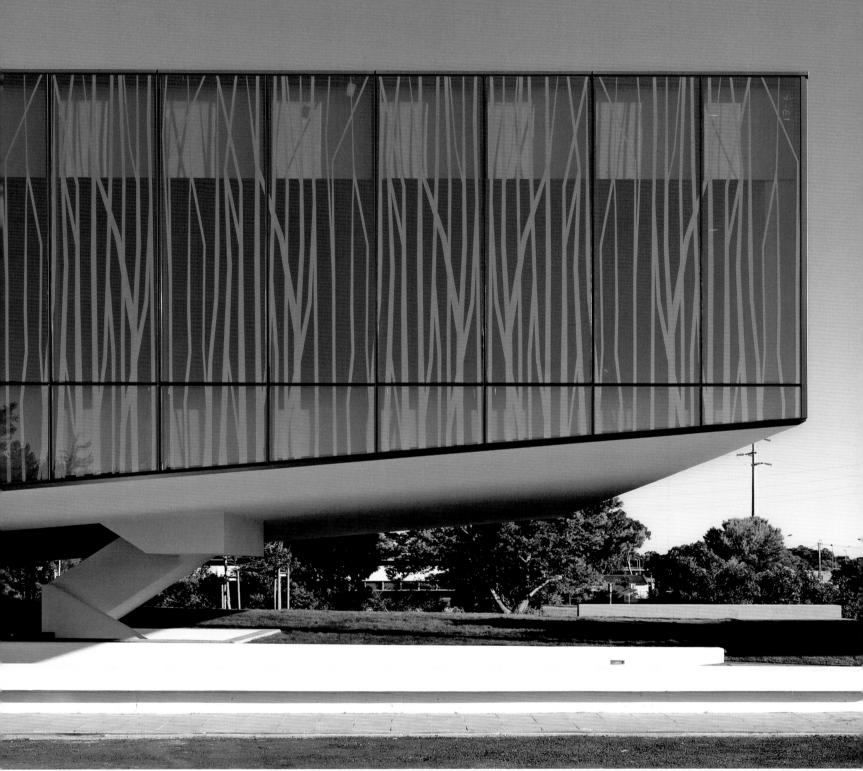

SHAPE ANALYSIS 造型分析

NEGATIVE SHAPE

Overall modeling shows its unique function through the "folding up" structure, and accommodate to the surrounding landscape to promote the harmonious coexistence between building and environment.

负造型

建筑整体构造独特鲜明,以"折叠"的构架方式很好地展示了建筑独有的功能,并与周围的环境紧密相连,取得了促进建筑与环境和谐共存的理想效果。

SOUTH ELEVATION

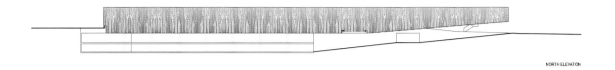

NORTH ELEVATION

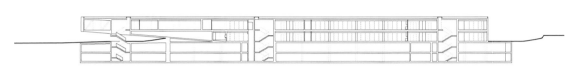

LONGITUDINAL SECTION

Elevation & Section
立面图和 剖面图

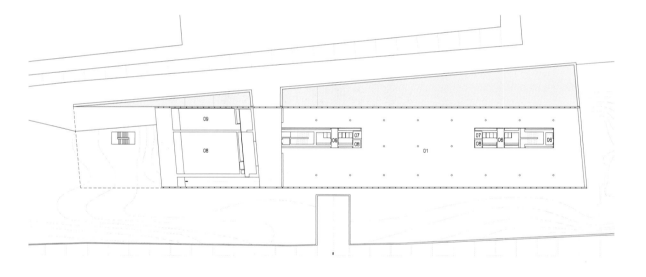

LEVEL 0

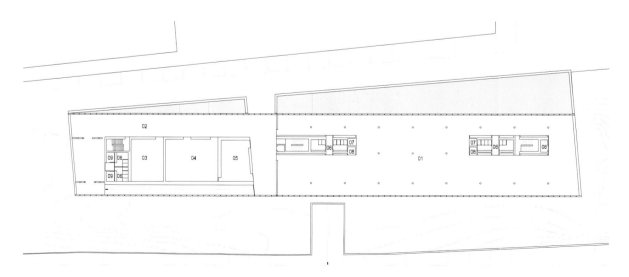

LEVEL 1

Floor Plan
楼层平面图

The rotation of the upper volume translates, on the base level, a negative that directly exposes the parking lot, allowing construction to mix with a green protection area. The building is intimately connected with its natural surroundings, taking advantage of it as a barrier against external agents.

In terms of materials, the solution is as restrained as possible to ensure the sobriety of the whole. Therefore, a limited range of materials is introduced in order to emphasize the formal clearance required: white spread surfaces and screen-printed glass are assumed as predominant materials.

位于底层的上层体块的转角转化为一个向里凹进的空间，直接通往停车场，使整个构造充分融入绿色保护区中。在这里，建筑与自然环境紧密相连，并充分利用它的屏障作用，使建筑远离尘嚣。

在材料的选用方面，设计方案尽量节俭，以确保整体结构的简练，展现出庄严肃穆的形象。为此，设计师限定了材料的使用量，以白色的延展型表面和丝网印花玻璃为主要材料，突出建筑的造型。

SKIN ANALYSIS 表皮分析

LIGHT SHAPE
White spread surfaces and screen-printed glass provide the building a piece of sobriety, while keep on its unique style.

轻表皮
白色的延展型表面以及丝网状印花玻璃设计，在保持自己独特风格的基础上，给建筑增添了一份庄严肃穆。

3M ITALY HEADQUARTER

3M 意大利总部

Architect: Mario Cucinella Architects
Location: Milan, Italy

设计公司：Mario Cucinella 建筑事务所
地点：意大利米兰

STRUCTURE AND MATERIAL 结构与材料

STRUCTURE
Steel Frame Structure
MATERIAL
Renewable Energy Material, Glass, Wood, Steel

结构
钢架结构
材料
再生能源、玻璃、木材、钢材

Sketch
概念草图

SHAPE ANALYSIS 造型分析

OPEN SHAPE

The project is located in the business park, building palisade structure is mainly composed of glass and wood structure, the transparent modelling introduces the surrounding beautiful environment into the building interior, also blend the building better into the surrounding environment.

开放式造型

本案坐落在商务公园内，建筑的围护结构主要由玻璃和木材构成。其通透的造型不仅使人们在建筑内部就能观赏到周边的优美环境引入建筑内部，还使建筑更好地融入周边环境中。

Site Plan
总平面图

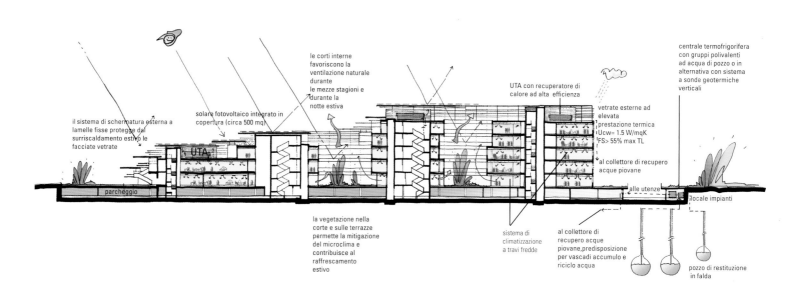

Summer Energy Strategies
建筑夏季能源运用图

The unique sinergy among 3M leadership in innovation and protection of the environment, Mario Cucinella Architects renomated vocation for sustainable design and Pirelli RE eco-building program has turned out into the first innovative building of the Malaspina Business Park located in the East outskirts of Milan.

The 11,300 m² and 5 stories building dynamic shape, the transparency enhanced by 2 internal courts, the natural colors integrated with green surroundings, the minimal energy needs optimized through brise-soleil and southern expo-sure, the intense use of renewable energy makes it the new architectural and green landmark of East Milano and perfectly match 3M's values of innovation, ethic in business, respect for employees and protection of the environment. The building is gas-free and all cooling and heating capacity is provided through a geothermal system where underground water is later released to the water ditches of "Parco della Besozza" to support its ecosystem. The reduced energy needs as well as geothermal and roof photovoltaic systems have allowed the building to get the Class A energy classification based on European EPBD (Energy Performance of Building Directive).

为了体现 3M 公司在创新和环境保护方面独一无二的领先地位，Mario Cucinella 建筑设计公司再次采用了可持续、生态型建筑设计，使该建筑成为米兰东郊马拉斯皮纳商务公园里的首个创新型生物建筑项目。

这栋五层楼高的建筑占地 11 300 m²，其立面造型动感十足，两个内场空间的设计更凸显了建筑的通透性，木色的整体外观与四周的环境浑然一体。此外，建筑的坐向朝南，采用了百叶窗设计，使建筑物的能耗降到最小。该建筑对再生能源的高效利用使其成为东米兰的新型绿色地标性建筑，极好地体现了 3M 的创新理念、商业道德、对员工的尊重和对环境的保护。建筑内部采用地热系统制冷和供暖，实现了建筑物的零排放。其地热系统的生态循环主要依靠地下水的排放和回流来支持。由于建筑的地热系统和屋顶太阳能系统在降低能耗方面的应用，该建筑荣获了欧洲建筑能耗指标 (EPBD) 的 A 级标准。

SKIN ANALYSIS 表皮分析

ENERGY SKIN

The initial design of the building is according to the sustainable requirements, in the skin design also use this concept, such as wood shading and barrier, the façade with shutter of the south side, roof solar system, etc.

节能表皮

建筑初始的设计就是按可持续发展的要求设计的,而这一理念在表皮的设计中也得到了很好的贯彻,如采用木材充当遮阳板与护栏、在南立面采用百叶窗设计、增设屋顶太阳能系统等。

SCIENCE PARK MECHATRONIK

科技园机电大楼

Architect: Caramel Architekten ZT gmbh
Landscape Architect: Idealice Alice Größinger
Client: BIG Bundesimmobiliengesellschaft m.b.H.
Location: Linz, Austria
Built Area: 72,312 m²
Photography: Hertha Hurnaus.

设计公司：Caramel Architekten ZT gmbh
景观设计公司：Idealice Alice Größinger
客户：BIG Bundesimmobiliengesellschaft m.b.H.
地点：奥地利林茨
建筑面积：72 312 m²
摄影： Hertha Hurnaus.

STRUCTURE AND MATERIAL 结构与材料

STRUCTURE
Steel Truss Frame
MATERIAL
Steel, Glass

结构
钢桁架结构
材料
钢材、玻璃

Site Plan
总平面图

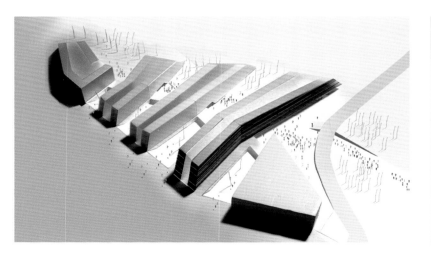

SHAPE ANALYSIS 造型分析

NEGATIVE SHAPE

To adapt to and take advantage of the wind while also echo with the surrounding landscape, the building takes a sloping shape to move part of the body into underground.

负造型

为了顺应风向、利用风能，以及与周围景观相呼应，本案采用斜坡式造型将一部分体量移至地下。

Model
模型图

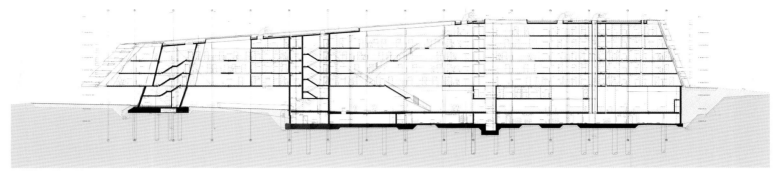

Cross Section 1
横剖面图一

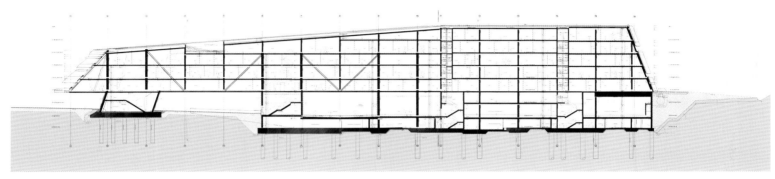

Cross Section 2
横剖面图二

The program involved designing several individual buildings which would be interwoven as well as tied to the existing University of Linz campus. The plan was to take into consideration the neighboring residential buildings as well as the natural form of the slope and the katabatic winds, which play an important role in keeping the city cool, and the poor condition of the building lot was not to be overlooked either.

The horizontal bending of the elongated blocks arose out of consideration to the existing structures. Moreover, the height of the building corresponds to the upper edge of the slope to the north and at the same time to the eaves of the residential buildings to the south. While In order to avoid an inflexible grid structure, straight lines were bent, thus also making full use of the grounds

本案由几个独立的建筑造型组合而成，它们彼此交织，与林茨大学的校园景观相互映衬。同时，本案的设计充分考虑到邻近住宅大楼的风格和北面的斜坡地形，以及如何引入凉爽的城市下吹风这三大问题。当然，施工过程中建筑用地将会发生的恶劣因素也被纳入考虑范围。

本案设计师考虑到现有的建筑结构，在狭长的建筑用地上设计出水平弯曲的造型。至于高度的设置也十分灵活，建筑北部不能高于斜坡，建筑南部要适应住宅大楼的屋檐。与此同时，为了避免网架结构过于僵化，设计师用曲线代替直线，同时也使得建筑用地被充分地利用起来。

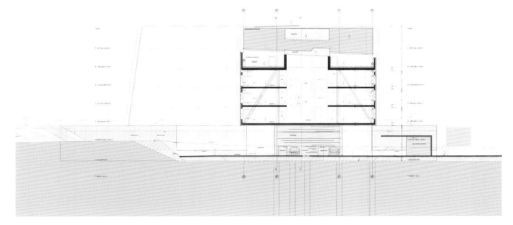

Cross Section 3
横剖面图三

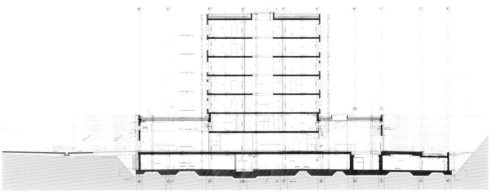

Cross Section 4
横剖面图四

SKIN ANALYSIS 表皮分析

ENERGY SKIN

The rails on the façade act as both a barrier of glass façade and shade from intense sun. It meets the need for daylighting and views and at the same time aviods direct solar radiation.
The design of the façade conforms to the structural system of the truss frame, cooperate with energy skin and the unique arrangement of barrier, all of these make the building crouch into the landscape.

节能表皮

立面的护栏设计兼有保护玻璃墙面和遮阳的作用，还能在满足采光和取景需求的同时，避免日光直射。这种运用节能表皮和特殊护栏的立面其实遵循了桁架的结构体系。所有的设计交融汇合才打造出如此雄浑壮阔的建筑奇观。

All in all, it was essential for the project to "crouch" into the landscape. This is why part of the slope was removed to make way for the basement, which houses the special rooms like laboratories and workshops and connects the individual building sections. Above this, hover the office wings between which the landscape flows into the grounds – a measure which was used to overcome the difficult task of tying the new structures to the existing campus, although these two areas are separated by a heavily used road (Altenbergerstraße). On the Science Park side, the grounds have been recessed enough so that one enters the underpass at ground level, following along a gradually sloping ramp that leads through the Park straight up to the buildings on the university campus side. In this way, the connection is at least in part at ground level. Above this level, as it were, extends the cantilevered structure of the first building element (Mechatronik), which forms the actual entranceway to the Science Park. Due to its enormous span and the deflection, the unit has been designed as a bridge structure, in which two massive cores support a 160-meter-long steel truss frame.

The design of the façade also conforms to the structural system of the truss frame. The parapets are not arranged randomly but have been placed to coincide with the points of greatest deflection. In this way the outward impression is diversified while the interior is marked by greater individuality.

总而言之，本案必须以"蹲伏"的结构形式融入周边景观。这就是为什么设计师要移除斜坡的一部分来打造地下室的原因。该地下室有一些特殊的作用，一方面用来设置实验室和工作间，另一方面用于连接某些独立建筑。地下室之上是办公大楼。办公大楼形似一对翅膀，中间被一条频繁使用的道路 (Altenbergerstraße) 隔开，让凉爽的下吹风得以顺畅地流动其中。这个"翅膀"设计出色地完成了为校园带来全新建筑造型这个艰难的任务。在建筑物一侧，有一个足够下凹的地面可以使人直接进入地下通道，你只需沿着一个渐斜的坡道走下去，就可以穿过这栋机电大楼抵达校园的其他大楼。这种楼与楼之间的连接方式，确保建筑至少有一部分是露出地面的。在地面上，这就好像将第一栋建筑（机电大楼）的悬臂式建筑结构扩展延伸，用于打造科技大楼的实际入口。由于跨度和挠度都极为之大，因此这个单元被设计成一个桥梁结构，由两个巨大的核心撑起了一个 160 m 长的钢桁架。

本案立面设计同样遵循钢桁架的结构体系，护栏的设计并非毫无根据，是根据建筑的每个选点的最大挠度设计而成的。这种做法，不仅让外观形象呈现出多种建筑结构，而且让内部设计也更个性化。

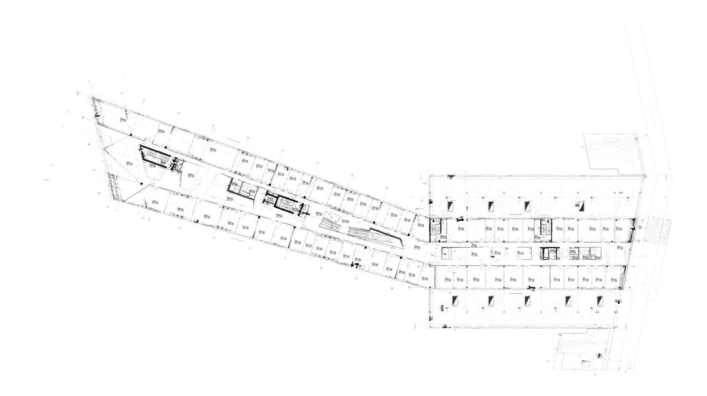

Floor Plan
楼层平面图

U15 办公楼

Architect: Cino Zucchi Architetti & General Planning
Client: Milanofiori 2000
Location: Milano, Italy
Site Area: 10,471 m²

设计公司：Cino Zucchi Architetti & General Planning
客户：Milanofiori 2000
地点：意大利米兰
占地面积：10 471 m²

STRUCTURE AND MATERIAL 结构与材料

STRUCTURE
Reinforced Concrete Structure
MATERIAL
Aluminium Sheet, Glass

结构
钢筋混凝土结构
材料
铝板、玻璃

SHAPE ANALYSIS 造型分析

PICTOGRAPHIC SHAPE

To meet the site planning, the building is designed into H-shape in order to maximize its scale. External flat part is be able to obtain sufficient light and ventilation, while the middle folded part is in favor of quiet and shady to meet needs of different users.

象形造型

为了配合场地的规划要求，建筑被设计成"H"形，以最大限度地利用场地。外部平坦的部分能获得充分的光照和通风，而中间的折叠部分则以营造安静和阴凉的环境为主，满足不同用户的需求。

The disposition on the ground and the typological layout of the U15 office building articulate and specify the general master plan guidelines in relationship to the specific features of the site and of the overall design of the open spaces. The large dimension of the building plot and of the maximum building envelope given by the master plan is interpreted by folding its perimeter inwards to generate a new layout which unites the advantages of an H-shaped scheme to that of a centralized one.

U15办公楼的外形和布局充分结合并突出了场地的总体规划及其开阔的空间特点。通过向内折叠外壳，整个建筑显示出了房屋地区图和总设计图中所规划的最大建筑外壳，形成了一种全新的布局设计，综合了"H"形结构布局和中央型布局两者的优势。

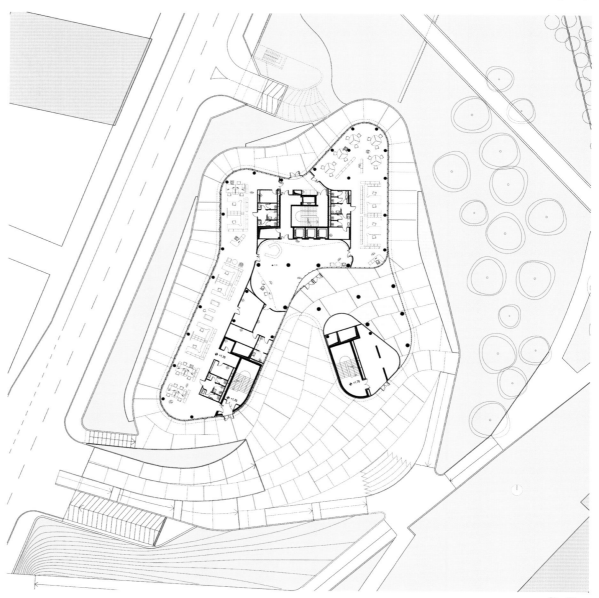

Site Plan
总平面图

SKIN ANALYSIS 表皮分析

ENERGY SKIN

The skin is set out carefully by aluminum plates in different colors and sizes that gives the building an outlook like a trunk. Aluminum plates not only avoid direct sunlight from east and west, but also gives a cooling effect to reduce temperature.

节能表皮

建筑的表皮主要由不同颜色、大小的铝板组成，塑造出建筑树干一般的外形。铝板的使用避免了建筑受到东、西面阳光的直射，同时还具有散热作用，起到了很好的降温效果。

The continuous perimeter "skin" which delimits the interior spaces is divided in several layers which donates depth and a strong light and shadow effect: the inner wall alternates a ribbon window band with an opaque one faced by anodized aluminum panels in natural warm grey, pale gold, light brown shades.

The exterior system of brise-soleils in aluminum sheets folded in different profiles, pierced by an array of custom-designed holes and treated with different anodized natural colors protects the windows from the east and west low sun transforming the building in a sort of large tree-trunk with an iridescent bark.

The design of the surrounding open spaces alternates paved and green surfaces; the former shape the paths reaching the main atrium from south and east, and offering themselves as a pleasant dehors for the ground floor offices.

建筑表皮在视觉上的变化强调了光与影的效果，界定了室内各个空间的布局，使其层次分明，呈现出更加开阔的视觉效果：内墙和外罩铝阳极氧化膜板的、不同色调（灰色、淡金色、浅棕色）的不透明墙板取代了带状窗。

按照不同剖面折叠的铝板遮阳窗外部系统由一系列自行设置的孔洞串联着，可以处理成不同的阳极颜色，带有闪光的色彩，使整个建筑物显得像大树干般，保护着建筑物不受东、西两边的太阳照射。

周围露天空间取代了铺地和绿化带的设计，使人们可以从南部和东部到达主厅，同时为一楼办公区提供了一个令人愉悦的室外空间。

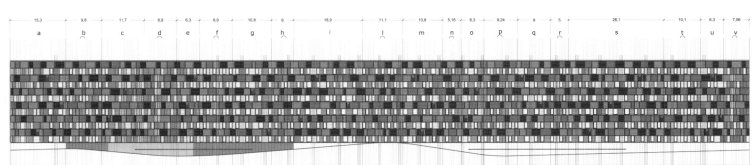

Elevation Sketch
立面概念图

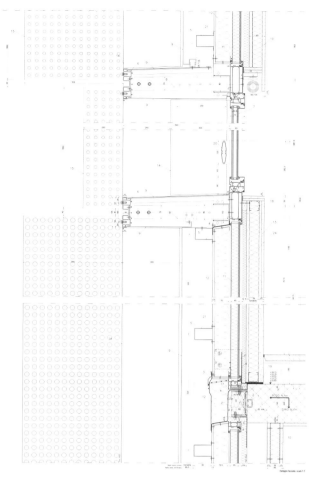

Detail Vertical Section 1
纵剖面节点图一

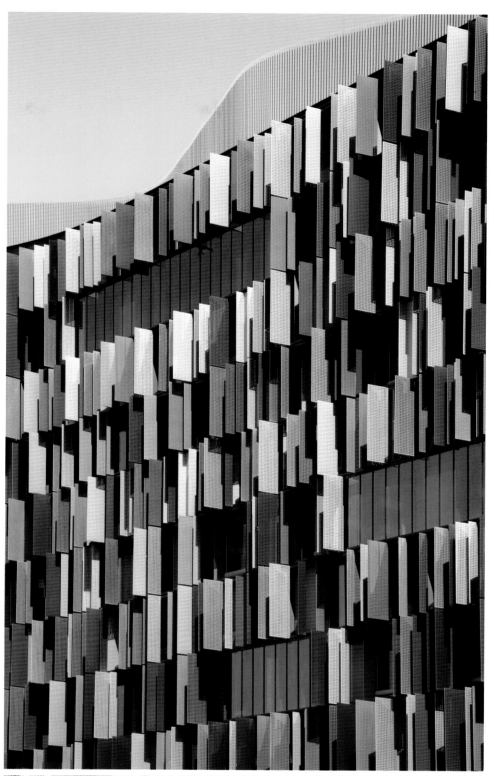

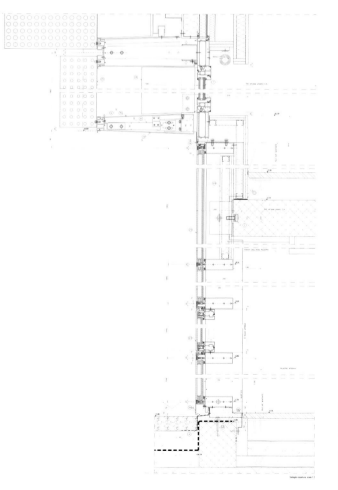

Detail Vertical Section 2
纵剖面节点图二

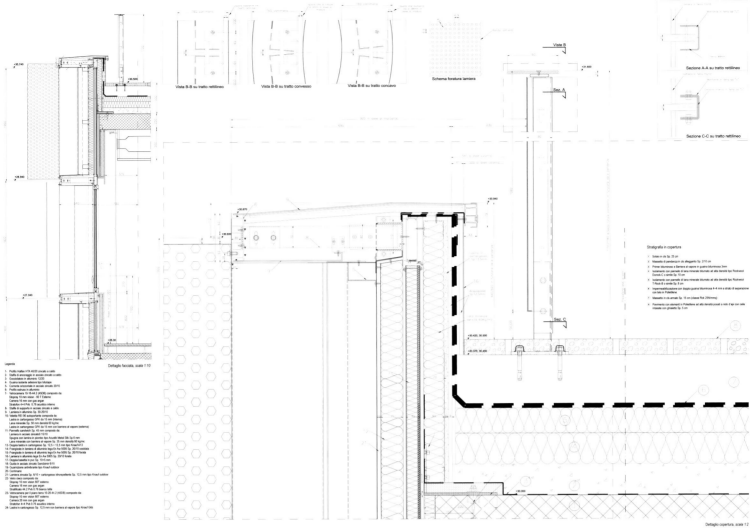

Roof Detail
屋顶细节图

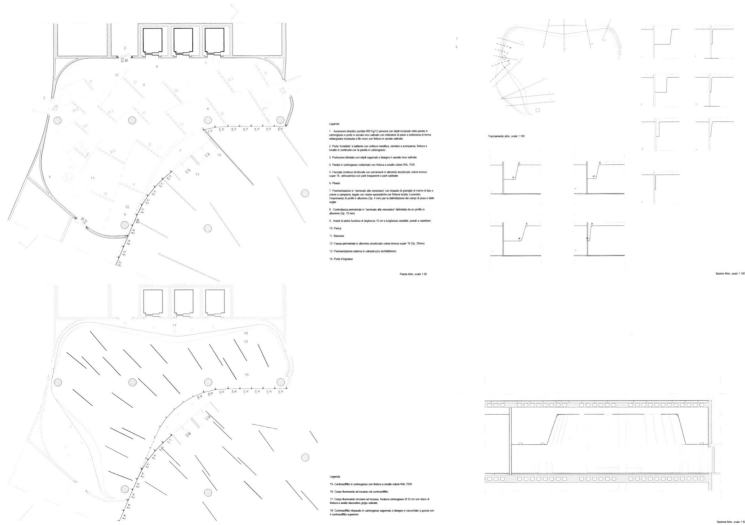

Hall Detail
大厅细节图

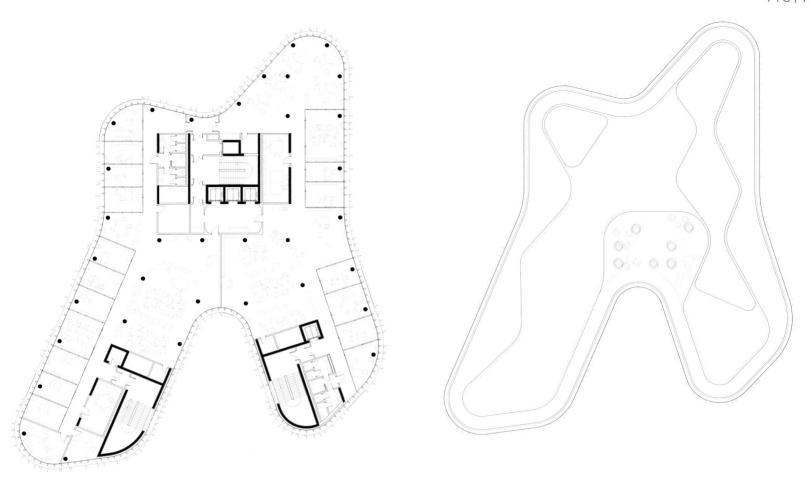

Type Floor Plan (Two Tenants)
典型楼层平面图（两户）

Roof Plan
屋顶平面图

OFFICE BUILDING IN PUJADES

西班牙 Pujades 办公楼

Architect: Josep Miàs
Client: PUJADES, 51-57 SL.
Location: Barcelona, Spain
Site Area: 7,000 m²
Photography: Jordi Bernadó

设计公司：Josep Miàs
客户：PUJADES, 51-57 SL.
地点：西班牙巴塞罗那
占地面积：7 000 m²
摄影：Jordi Bernadó

STRUCTURE AND MATERIAL 结构与材料

STRUCTURE
Steel Frame Structure
MATERIAL
Glass, Steel

结构
钢架结构
材料
玻璃、钢材

The building lot is situated between three streets: Pujades, Pamplona and Pere IV (one of the street that doesn't respond to the grid of Cerda's Plan because it's old condition as an historic road). Certainly, it is an irregular plot where the traces of the old division in plots system in relation to Pere IV Street are still maintained. This geometry allows setting up the project in relation to the existing buildings that are preserved, such as the residential buildings on one side of the block and the old industrial building that faces Pamplona Street on the other side, a brick and stone construction in which public equipment will be settled.

The aim of the project is to rebuild from the old plot division rules, following a geometry that obeys to Pere IV Street orientation. On the other hand, the irregular geometry of the plot brings the opportunity to walk across the project from one street to the other. Somehow, it is an act that makes losing sense to the block criteria of Ildefons Cerdà. The building organizes itself from an optimal distance between blocks in order to locate offices and the result is the creation of three main buildings crossing across from Pujades Street to Pere IV. The elevations of the building show clearly how these transversal secondary streets appear between the three structures, as if they were passages, and allow to illuminate the interior spaces of the offices.

Consequently, the geometry does not maintain a constant alignment. It presents some doubts about the section in relation to the passage or the street-courtyard and in relation to the existing industrial building in Pamplona Street as well. That's the reason why the building reacts and overflies the old construction on the upper floors. The project in 22@ presents a fragmented building with vertical voids between the volumes. The spatial complexity is broken in a middle point between the main streets and a transversal axe generates all vertical accesses with stairs, elevators and vertical installations with pipes and metallic ducts.

Underground, the building presents a two-floor parking with access next to the preserved industrial building. The structural system is a steel skeleton that presents several levels of complexity. The building is proposed as six independent entities joined by a transversal central passage where stairs and elevators are located. Each one of the six entities is composed by multiple bracings with diagonals or steel crosses which make it stable on its own.

Finally, the higher building exhibits a trapezoidal cantilever on the two upper floors, as if it was trying to cover the neighbour building. Steel diagonals are used to hold the structure and reproduce a huge beam of seven meters tall by sixteen meters long.

Location
区位图

SHAPE ANALYSIS 造型分析

OPEN SHAPE

The project must be built on the basis of preserving the historical structure alterations, but also reflecting the sense of modern. These two elements combined produce this transparent building in the form of stacked "square box" with open structure so that visitors will feel closer to the combination of modern and historical.

开放式造型

本案忠于原有结构的前提下进行改建。它要求既不触动传统结构造型，又得充分体现代摩登风格。现代摩登与历史造型交融汇合，打造出仿佛用很多"方盒子"堆积而成的半透明大楼。经试验，游客似乎更愿意接受这种崭新的风格与造型。

本案的规划用地处于三街交界处：Pujades 街、Pamplona 街和 Pere IV 街（由于其中一条街为历史名街，因此未被列入 Cerda 关于网格设计的规划中）。虽然 Pere IV 街要被征用土地，但是相关的历史遗迹依然被保留。本案的规划方案与已有建筑物的保留项目相关，例如街道一侧的住宅小区，街道另一侧朝向 Pamplona 街的工业建筑，乃至那个砖石搭建的公用设备也将被保留。

本案在遵循原有的建筑架构的前提下进行改造，并且新式的几何结构必须按照 Pere IV 街的走向。根据 Pere IV 街的走向来设计这个不规则的几何结构，主要想实现街道与街道之间的相互连通，让人们可以通过建筑从一条街走向另一条街。从某种程度上来评定，这种设计不符合 Ildefons Cerdà 的社区标准。虽然本案的主楼建筑与社区其他楼保持适当距离，但是为了规划更便利的办公写字楼，不得不打破社区的标准——让三个主楼建筑从 Pujades 街横跨至 Pere IV 街。从建筑物三面图上就能清晰地看到，三条街道好像人行过道一样横跨三个主楼建筑。归根结底，这是为了办公室区域引进更充足的光线而特别设计的。

不过，本案的几何结构不是对齐的形式。关于保留 Pamplona 街内通道、庭院过道和 Pamplona 街的工业建筑的设计方案在项目实施前备受质疑。其实 22@ 项目给本案以很多启发，它成功塑造出带有垂直孔洞的"支离破碎"建筑。因此，本案借鉴了 22@ 的项目设计，将所有建筑视为一个空间综合体，在主街道之间取中点，将中点当横轴构造。横轴上分布着楼梯、电梯、管道、金属管垂直装置，这种设计使得多种建筑空间有了连接。本案出色地完成了目标，它保留了旧建筑，打造出与上层旧建筑形成对立又超越了旧建筑的新式建筑。

本案地下有两层楼停车场，停车场入口紧挨原来的工业建筑大楼。结构体系采用钢骨架，能呈现出多层复杂样式。本案可分为六个独立的主体，这六个主体都靠近一条横向延伸的主干通道，通过这条主干通道便到达各个楼梯和电梯。每个主体由多种带有对角支柱或十字钢架的材料支撑着，使得主体结构十分坚固。

最后，稍高的那栋建筑在两层高的位置使用对角钢支柱支撑起整个建筑物的梯形悬臂。这个高 7 m、长 16 m 的"巨型横梁"——梯形悬臂，远远看去，好像要将相邻建筑都覆盖了一样。

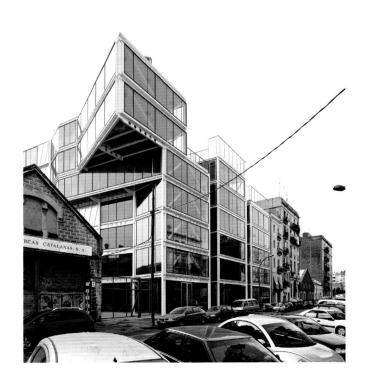

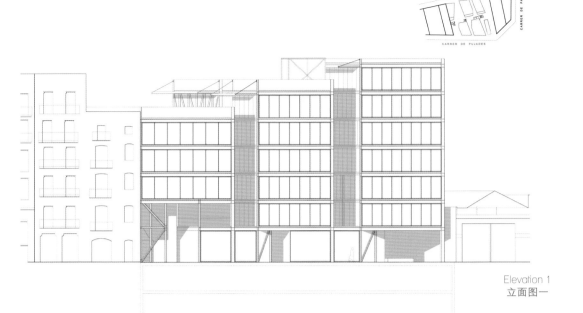

Elevation 1
立面图一

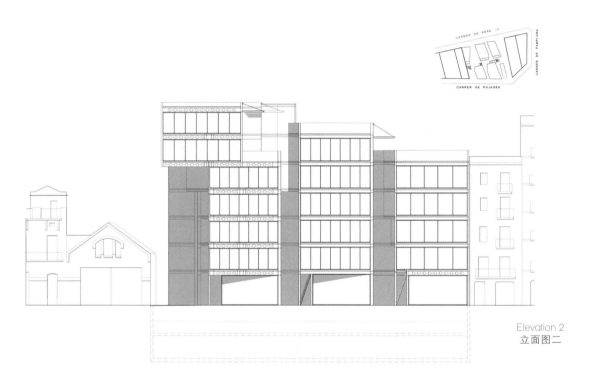

Elevation 2
立面图二

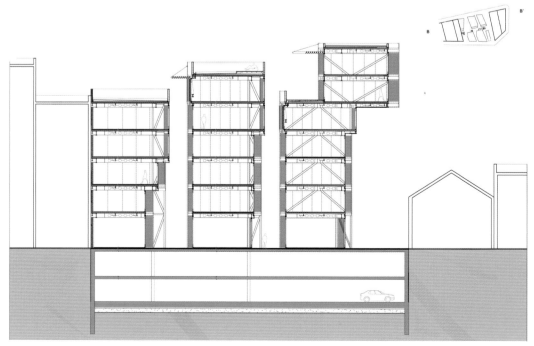

B-B Sectional
B-B 剖面图

SKIN ANALYSIS 表皮分析

TRANSPARENT SKIN

Glass façade on full range ensures that each unit in any period of time, in any orientation will enjoy plenty of sunlight and also reduce the sense of estrangement of the buildings with neighborhoods.

透明表皮

全方位的玻璃立面保证各单元在任何时段、任何方位都能享受到充足的阳光，此外还能减少建筑和街区的隔阂感。

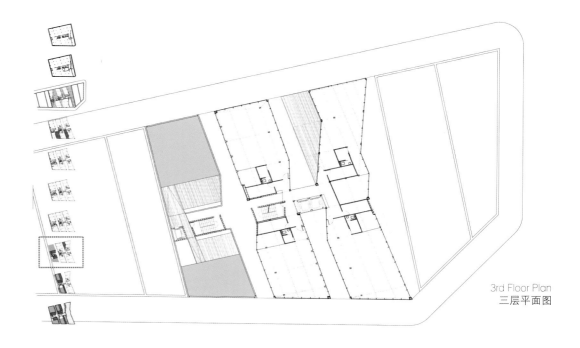

3rd Floor Plan
三层平面图

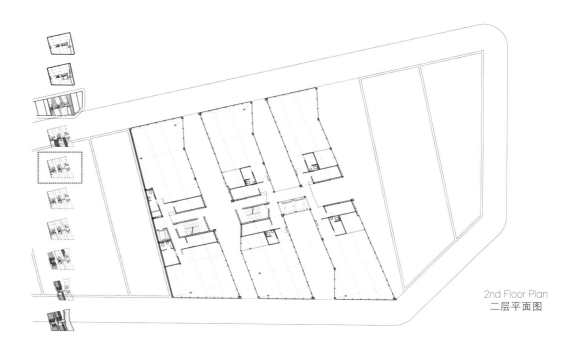

2nd Floor Plan
二层平面图

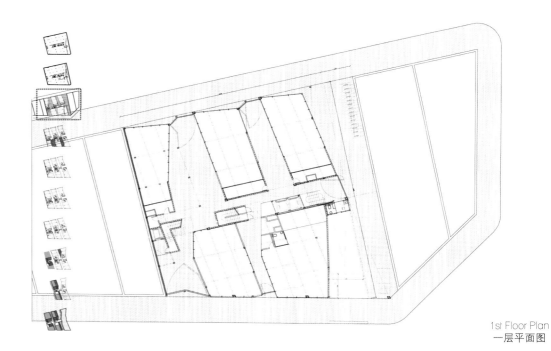

1st Floor Plan
一层平面图

BOLOGNA CIVIC OFFICES

博洛尼亚市政办公楼

Architect: Mario Cucinella Architects
Location: Bologna, Italy
Site Area: 33,000 m^2
Photography: Daniele Domenicali

设计公司：Mario Cucinella Architects
地点： 意大利博洛尼亚市
占地面积：33 000 m^2
摄影：Daniele Domenicali

STRUCTURE AND MATERIAL 结构与材料

STRUCTURE
Reinforced Concrete Structure

MATERIAL
Steel, Concrete, Glass

结构
钢筋混凝土结构

材料
钢筋、混凝土、玻璃

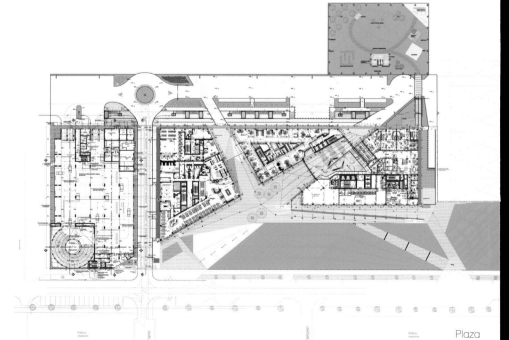

Plaza
广场平面图

Bologna City Council commissioned a new 33,000 m² office building to house their Municipal head offices. They needed to bring together 1,000 employees currently scattered in 21 locations throughout the city.

The new building is situated behind the central train station in the Bolognina district on the site of the former fruit and vegetable wholesale market. The project seeks to upgrade the area and re-connect it to the city centre.

博洛尼亚市议会委托建造一座占地 33 000 m² 的办公大楼，作为其总部办公楼。他们希望新办公楼能满足 1000 多名员工的办公需要，能将散布在城市各处的 21 个分部凝聚在一起。

本案位于 Bolognina 区的中央火车站后方，这里曾经是水果蔬菜的批发市场。本案旨在增创区域发展优势，与市中心重新建立联系。

SHAPE ANALYSIS 造型分析

OPEN SHAPE

The design concept is to break a single mass into three distinct blocks destined for different activities. The three blocks of different heights are unified by a folding shading roof, a four storey atrium entrance and a new sloping public space.

开放造型

玻璃幕墙使本案外观看起来通透豁亮，折叠式遮阳屋顶既防日光辐射，又令三座独立体量凝聚起来，发挥着双重功能。

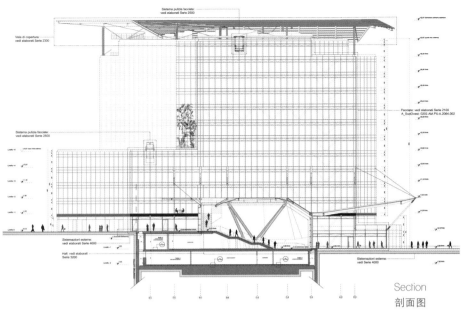

Section
剖面图

SKIN ANALYSIS 表皮分析

TRANSPARENT SKIN

Glazed façade offers a transparent and bright appearance for the building. The shading roof serves the dual purpose of providing solar protection and lending architectural cohesion to the complex.

透明表皮

本案试图打破体量单一化的传统设计理念，打造三座相对独立而功能各异的体量。虽然这三座体量高度各不相同，看似相对独立，实则被巨大的折叠式遮阳屋顶、四层高的中庭入口和崭新的倾斜的公共空间关联起来。

This large single cover is the defining element of the design. It is folded like a giant "origami" that rests gently on the various buildings above a panoramic terrace.

The programme includes space for shops, offices, services and sports facilities.

这片巨大的屋顶是本案的标志性元素。它就像一张巨大的"折纸"，轻轻地覆盖在三座高低不一的体量上，打造出一个全景式露台。

本案设有商铺、写字楼、服务区和运动场所等。

Sketch
概念草图

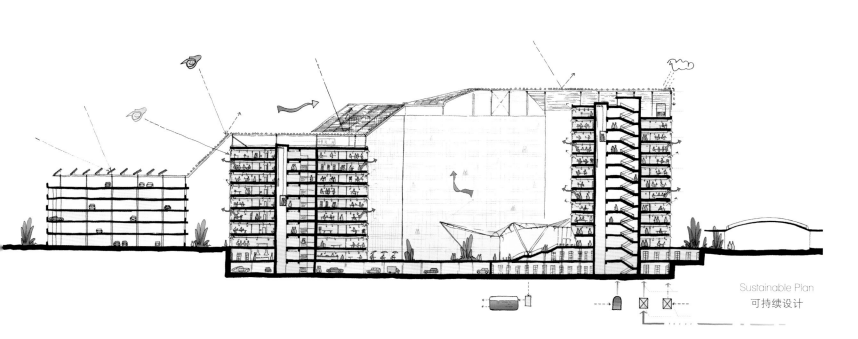

Sustainable Plan
可持续设计

CIVICISM
市政

NOAIN CITY HALL
Noain 市政厅

Architect: Zon-e Arquitectos
Client: Noain City Council
Location: Navarra, Spain
Photography: Pedro Pegenaute

设计公司：Zon-e 建筑事务所
客户：Noain 市议会
地点：西班牙纳瓦拉
摄影：Pedro Pegenaute

STRUCTURE AND MATERIAL 结构与材料

STRUCTURE
Steel Frame Structure
MATERIAL
Steel, Metal, Glass

结构
钢架结构
材料
钢材、金属、玻璃

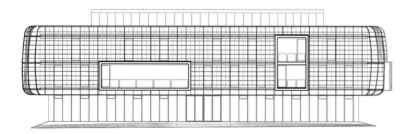

Northeast Elevation
东北立面图

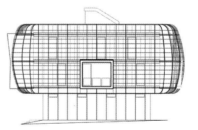

Northwest Elevation
西北立面图

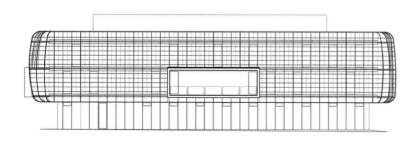

Southwest Elevation
西南立面图

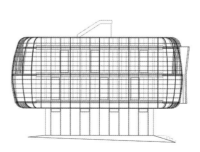

Southeast Elevation
东南立面图

SHAPE ANALYSIS 造型分析

DIGITAL SHAPE

The new building plays an important role in coordinating the open plazas and parks. Grid and could - like Patterns are all completed by computer simulation and design.

数字造型

新市政厅大楼在协调开放式广场和公园的关系中起着重要作用。本案中所使用的网格形态和云状图案均是借助计算机虚拟技术完成的。

Traditionally, the City Hall had a noticeable institutional character and its presence was severe and somewhat hieratic. As time went by, it began to lodge other functions and become a place of meeting for the citizens. As a result, its image became smoother, and thus the architecture reacted becoming more pleasant. Now, in the beginning of the XXI century, the City Hall is transforming into a model of interaction with the environment by implementing mechanisms in the face of increasing energetic challenges that our society can no longer deny.

The external membrane is a metallic latticework with an organic form where vegetation will grow producing a "cloud" that will change its density and colors throughout the year.

The vegetal canvas is composed by the Virginia creeper (Veitchii), which climbs and covers the façade during the summer, protecting the building from the solar radiation and also serving as a refreshing device. During the fall, it acquires spectacular red shades and in the winter, due to its deciduous leafs, lets sunlight pass and heats the double inner skin.

After a rigorous study of the energetic behavior with powerful digital analytical tools, the building received an outstanding A level qualification with a reduction of power consumption of 60%. Thus the building will be perceived as a landscape that changes, displaying the course of days and seasons. In summary, it acts as an index of the everyday life and the yearly cycles of the Noain citizens.

以前的市政厅固守旧制，有明显的等级之别，建筑风格古板苛刻兼有几分拒人于千里之外的神圣威严。随着时代的发展，随着市政厅新增了不少职能，市政厅俨然成为吸引公民参政议政的特别场所。随之而来的，政府的形象亦变得圆滑世故，其建筑外观也相应地呈现出一种平易近人的姿态。二十一世纪初，为了迎接更多新鲜的挑战，为了让社会大众更喜欢市政厅，本案根据相关实施机制，以打造环保建筑典范为目标，对原市政厅进行了改造。

本案的外层是一个有机金属网格薄膜。这种金属网格能栽种植物，植物生长后将形成一种"云状"图案，能随年月而变更密度和颜色。

本案外表面被清爽提神的绿色植物——爬山虎所覆盖。它旺盛的枝叶在建筑表面肆意攀爬，能为人们阻挡盛夏里猛烈的日光；红艳艳的枝叶展现出一派蓬勃生机，能为人们驱走冷秋里的沉寂；当它卸掉枯枝败叶，还能让建筑外表皮吸收阳光热量，为人们带来冬日和暖的气息。

本案在一次采用极为严格的数字分析工具进行调研的评比中，因减少了高达 60% 的电力消耗而荣获 A 级环保认证。由于建筑物能自发地改变外表形态，呈现出晨昏之别、季度相异的极致美景，所以本案被业内视为优秀而成功的范例。无论如何，这栋建筑已然渗入到 NOAIN 市民的日常生活中，成为最直观的"时钟"。

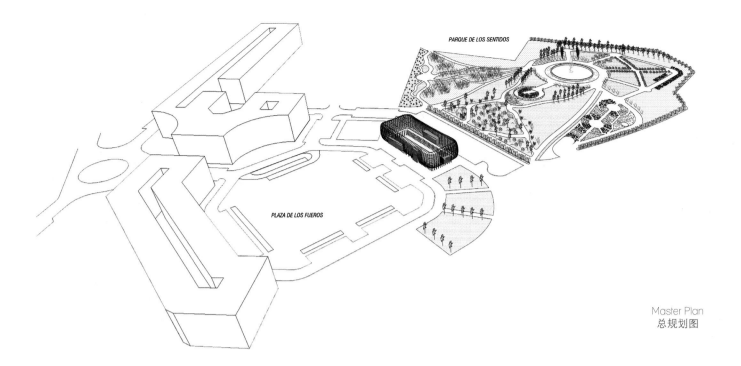

Master Plan
总规划图

Base geometry

Modeling traces

Supporting beams

Estructural units

Vertical frame

Horizontal frame

Estructural whole

'Green growing grid'

Volume Structure
体量结构图

ENERGÍA

A. Células fotovoltaicas
B. Vebtilación mecánica
C. Suelo radiante
D. Captación geotérmica

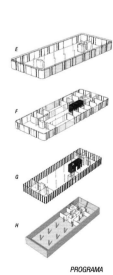

PROGRAMA

E. planta diáfana
F. planta privada
G. planta pública
H. planta de servicio

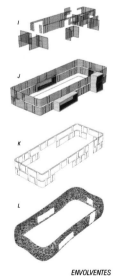

ENVOLVENTES

I. particiones interiores de vidrio
J. piel interior vidrio traslúcido
K. piel exterior policarbonato
L. 'nube vegetal'

Interior Structure
内部结构图

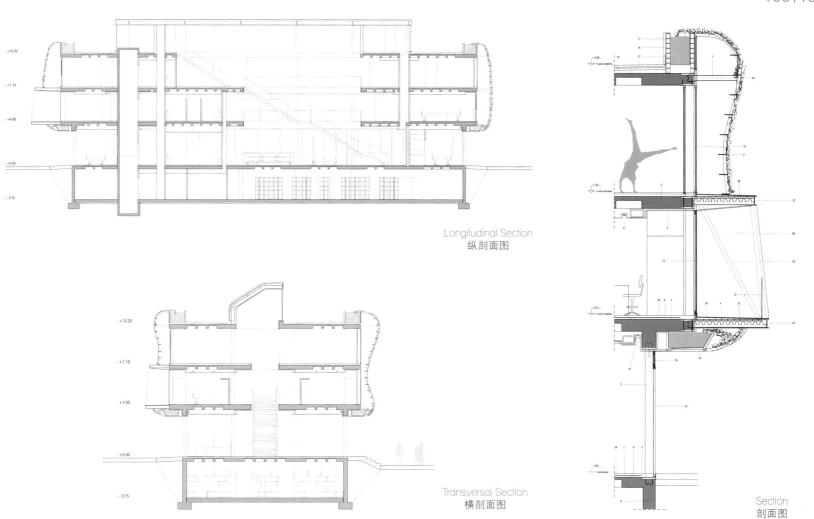

Longitudinal Section
纵剖面图

Transversal Section
横剖面图

Section
剖面图

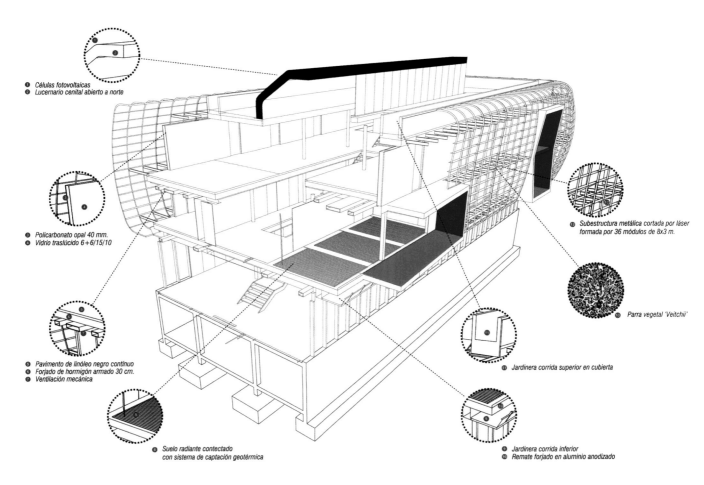

① Células fotovoltaicas
② Lucernario cenital abierto a norte
③ Policarbonato opal 40 mm.
④ Vidrio traslúcido 6+6/15/10
⑤ Pavimento de linóleo negro contínuo
⑥ Forjado de hormigón armado 30 cm.
⑦ Ventilación mecánica
⑧ Suelo radiante conectado con sistema de captación geotérmica
⑨ Subestructura metálica cortada por láser formada por 36 módulos de 8x3 m.
⑩ Parra vegetal 'Veitchii'
⑪ Jardinera corrida superior en cubierta
⑫ Jardinera corrida inferior
⑬ Remate forjado en aluminio anodizado

Details
节点图

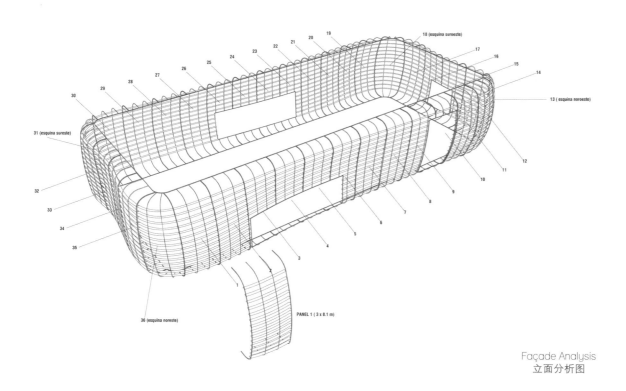

Façade Analysis
立面分析图

SKIN ANALYSIS 表皮分析

ECOLOGICAl SKIN

The building uses the organic metallic latticework as the external membrane, with Virginia creeper cool in the summer while protect the building from the solar radiation in winter. The virginia creeper's vicissitude keep building exterior changed continuously, even it is a good green decoration.

生态表皮

本案使用有机金属网格作为建筑外表皮，同时在建筑表面种植爬山虎。爬山虎具有多种作用：其一，它能起到夏季降温的作用；其二，它能阻挡冬日的辐射；其三，它的枯荣转变会直接改变建筑外观，使建筑呈现不同的外表面；其四，它还是天然养眼的绿色屏障，能舒缓视觉疲劳。

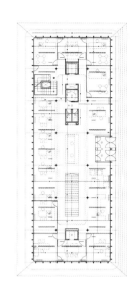

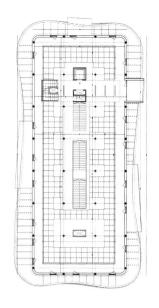
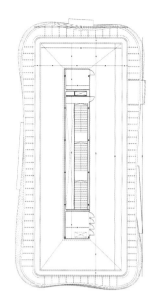

Floor Plan 1
楼层平面图一

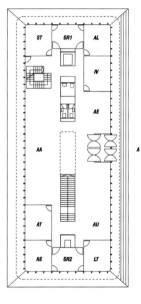
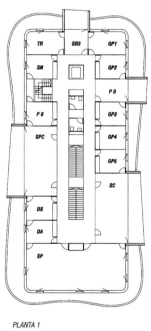
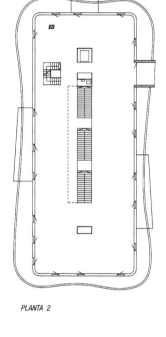
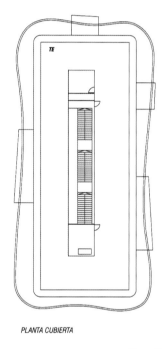

PLANTA 0　　　　　PLANTA 1　　　　　PLANTA 2　　　　　PLANTA CUBIERTA

Floor Plan 2
楼层平面图二

HASSELT COURT OF JUSTICE HASSELT BELGIUM

比利时哈瑟尔特市法院

Architect: J. MAYER H. Architects, a2o-architecten, Lenssass architecten
Client: Stationsomgeving NV Hasselt
Location: Hasselt, Belgium
Site Area: 3,640 m²
Photography: Filip Dujardin

设计公司：J. MAYER H. 建筑事务所、a2o-architecten、Lenssass 建筑事务所
客户：Stationsomgeving NV Hasselt
地点：比利时哈瑟尔特市
占地面积：3 640 m²
摄影：Filip Dujardin

STRUCTURE AND MATERIAL 结构与材料

STRUCTURE
Reinforced Concrete Structure

MATERIAL
Steel, Concrete, Timber

结构
钢筋混凝土结构

材料
钢材、混凝土、木材

The Court of Justice is one of two iconic projects within the new urban development around the main rail station. The logistics and siting of a courthouse with multiple security barriers of security results in a massing composed of three interconnected volumes. References include the old industrial steel structures that formerly occupied and defined the site, their organic Belgian Art Nouveau forms constituting part of the cultural heritage of Hasselt.

该法院是新城市围绕其火车站而发展的两个标志性项目之一。综合其物流、选址和多重安全壁垒三个因素，整个建筑由三个相互关联的体块组成。建筑的设计参照了传统工业中的钢结构，这在建筑中占据了重要位置，并起到了决定性的作用，结构中蕴含的比利时新艺术形式成为哈瑟特尔文化遗产的重要组成部分。

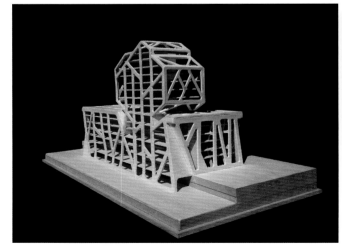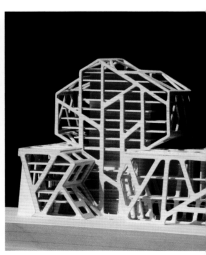

Model
模型图

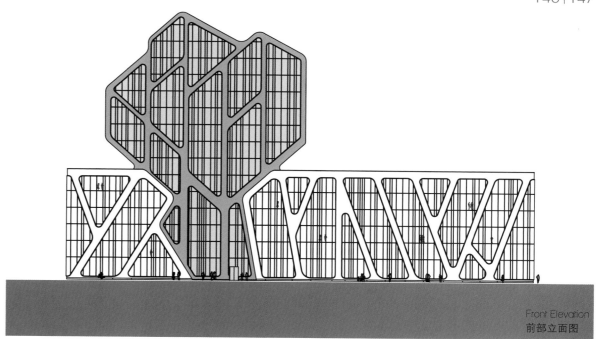

Front Elevation
前部立面图

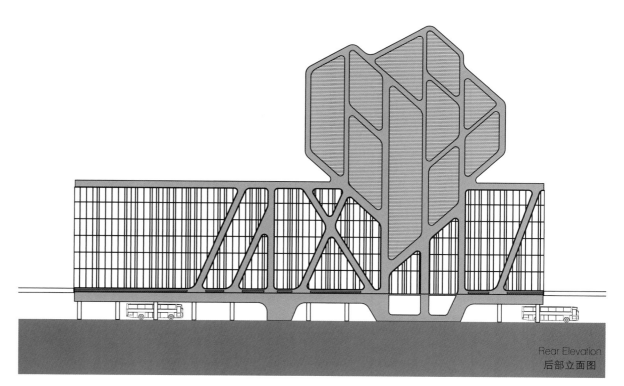

Rear Elevation
后部立面图

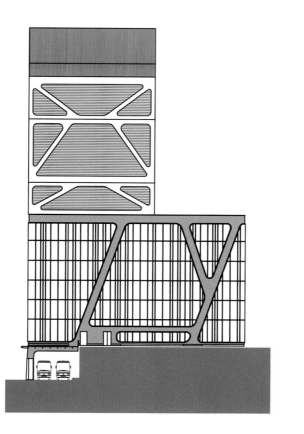
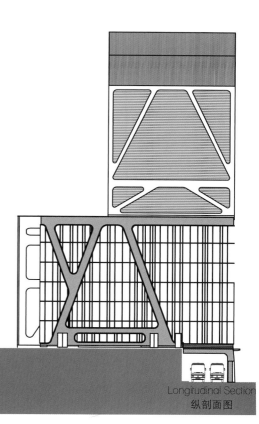

Longitudinal Section
纵剖面图

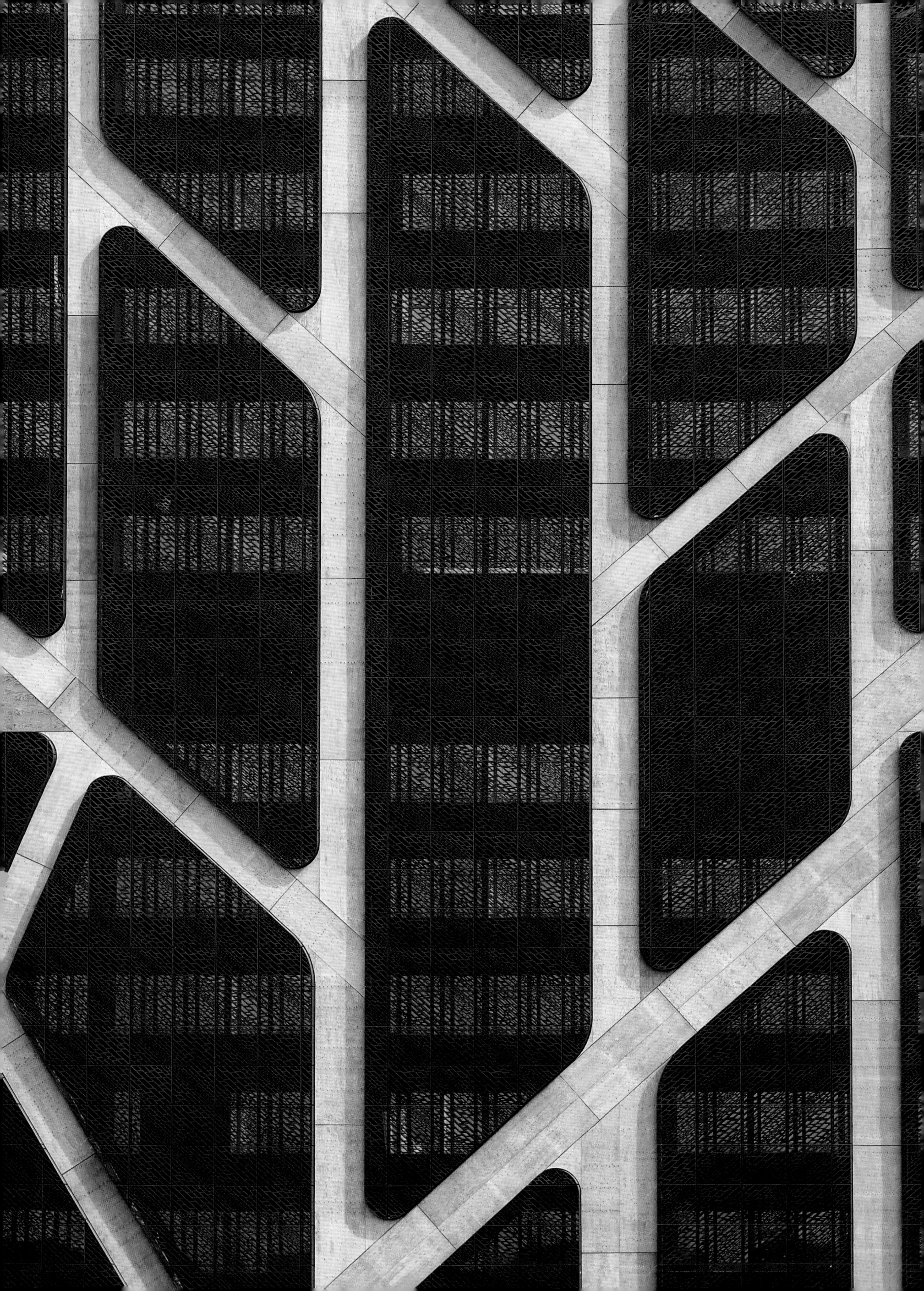

SHAPE ANALYSIS 造型分析

PICTOGRAPHIC SHAPE

Shapes echoes of a tree, which, in addition to being is the Hasselt town emblem, as well as also harks back to the historicpre-medieval European tradition of holding a special "place of speaking justice" underneath a large tree in the center of a dwelling.

象形造型

建筑的体量和"树"的外形相呼应,这不仅是哈瑟尔特镇的象征,也反映了欧洲中世纪时期,人们在城镇中心的大树下进行"公正的会谈"的传统。

SKIN ANALYSIS 表皮分析

LIGHT SKIN

The building combined with evey features of the cultural heritage of Hasselt, depand on traditional steel structure and light skin to become a kind of landmark construct.

轻表皮

建筑融合了哈瑟特尔文化遗产中所有的特色,依照传统的钢结构建设完成,配合轻表皮的运用,打造了一座标志性建筑。

RHÔNE – ALPES COUNTY COUNCIL HALL

法国罗纳 – 阿尔卑斯大区市政厅

Architect: Christian de Portzamparc
Client: Regional Council Rhône-Alpes
Location: Lyon, France
Site Area: 45,650 m²
Photography: Gitty Darugar, Nicolas Borel
Drawings: ACDP

设计公司：Christian de Portzamparc
客户：罗纳·阿尔卑斯大区区市政厅
地点：法国里昂市
占地面积：45 650 m²
摄影：Gitty Darugar、Nicolas Borel
图纸：ACDP

STRUCTURE AND MATERIAL 结构与材料

STRUCTURE
Concrete Beam Structure

MATERIAL
Glass, Wood, Concrete

结构
混凝土梁柱结构

材料
玻璃、木材、混凝土

As the French regions grow in power and influence, it is increasingly important for that power to be present and symbolized in urban spaces. But how do you go about turning a 40,000 m² office complex into a symbol and a public space where everyone can meet? This building presents a heart, a visible interior, made up of discussion spaces and meeting points leading to a large council chamber. This allows for a flow of natural light and provides views from multiple angles. To get from one department to another, you need to cross an interior landscape. This new type of open-block building is open to the city, with an interplay of empty and full spaces creating a visible forum, a political "square", a large and accessible city hall.

随着法国地方分权制度的改革，地方政府的权力与影响力日渐扩大，运用地方权力与建设城市地标显得日益重要。但是如何将40 000 m² 的综合办公楼打造成一座能满足广大群众休闲活动需求的地标性建筑呢？本案设计出一个围绕中心、四面通透的建筑。分布各个方位的讨论室与会客房彼此连通，围绕"中心"大型会议厅而建。通透的室内环境不仅便于引入源源不断的自然光线，还方便游客多角度观景。当你从一个部门走去另一个部门时，沿路还能欣赏到各种室内景观。这种新型的开放式建筑面向所有市民开放。架空层与全空间的巧妙搭配，创造出一个透明度极好的公共座谈场所。它不仅仅是一个公正严明的政治广场，还是一座亲近民众的市政厅。

SHAPE ANALYSIS 造型分析

OPEN SHAPE

Though the appearance appeals strong, the building has an internal hollowed structure. The atrium set aside large open free spaces to facilitate free access for relevant personnel, and increased transparency in government office.

开放式造型

尽管本案外观给人厚重沉稳的感觉，但是其内部却采用了空心设计。本案将中庭划分出多个开放自由式空间。这些无所阻隔的自由空间，一方面方便相关人员自由出入，另一方面大大提高了政府的办公透明度。

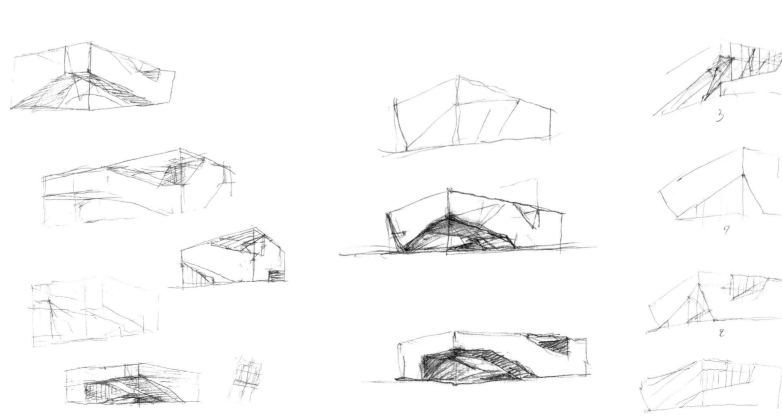

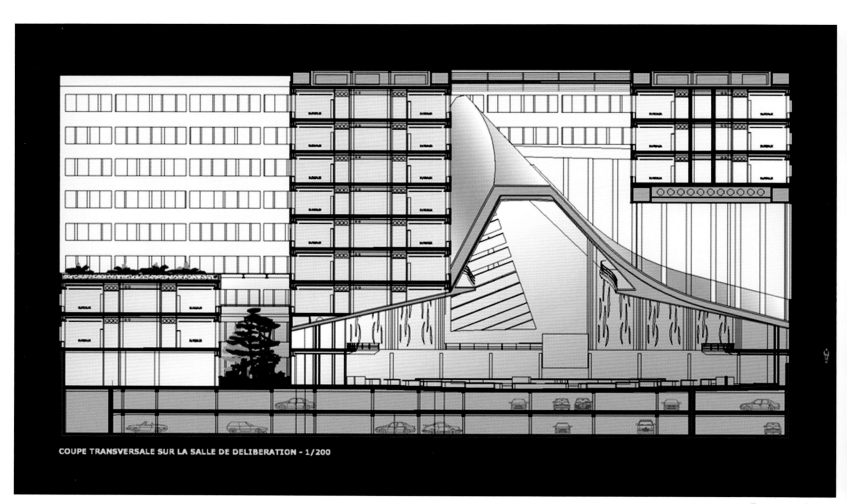

COUPE TRANSVERSALE SUR LA SALLE DE DELIBERATION - 1/200

Transversal Section
横截面图

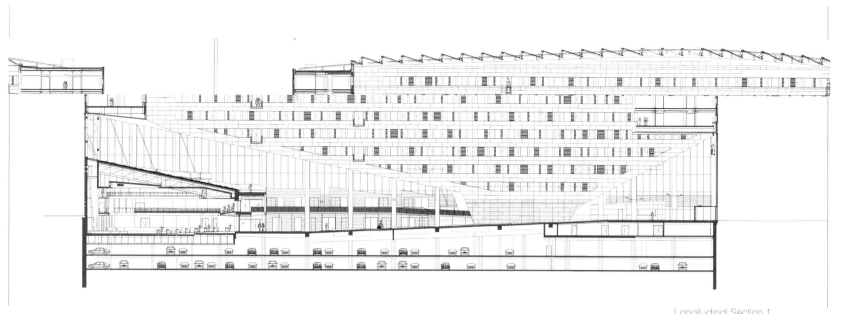

Longitudinal Section 1
纵剖面图一

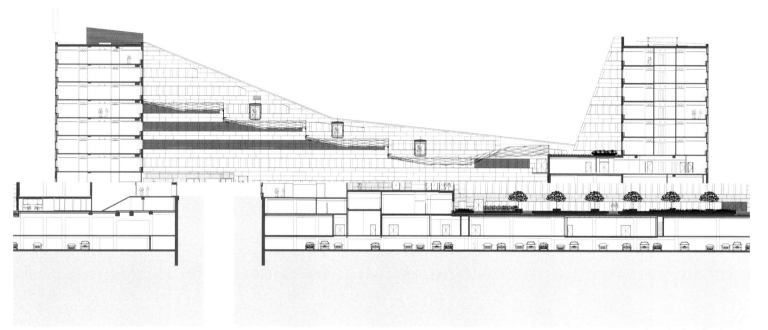

Longitudinal Section 2
纵剖面图二

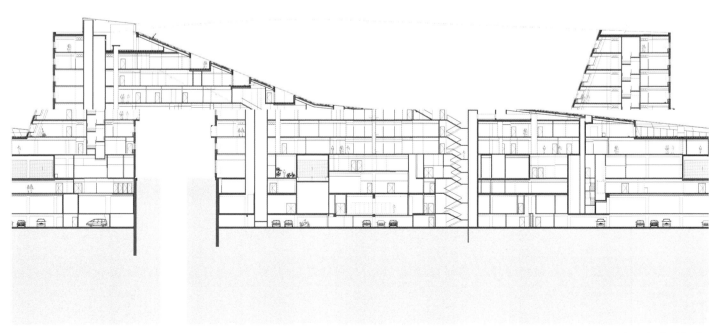

Longitudinal Section 3
纵剖面图三

SKIN ANALYSIS 表皮分析

HEAVY SKIN
Founder shape and heavy stone façade expresses the authority and solemn of government departments, but also reflects history and culture.

重表皮
厚重的石材立面打造出规则方形形态,既突出了政府部门的权威与庄严,也体现出当地浓厚的历史文化气息。

SUN MOON LAKE ADMINISTRATION OFFICE OF TOURISM BUREAU

日月潭风景管理处

Architect: Norihiko Dan Architects
Location: Taiwan, China
Built Area: 33,340 m²

设计公司：Norihiko Dan 建筑事务所
地点：中国台湾
建筑面积：33 340 m²

STRUCTURE AND MATERIAL 结构与材料

STRUCTURE
Reinforced Concrete Structure
MATERIAL
Concrete, Glass, Steel

结构
钢筋混凝土结构
材料
混凝土、玻璃、钢材

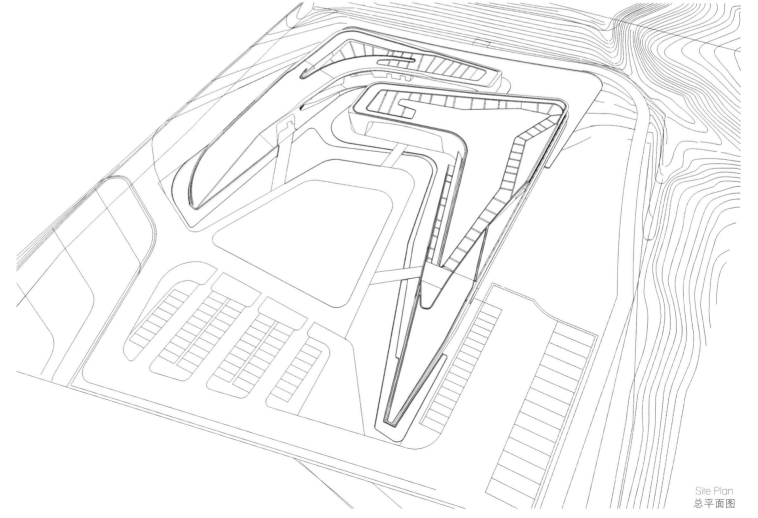

Site Plan
总平面图

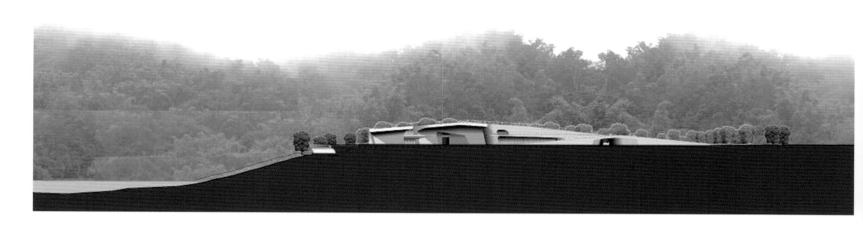

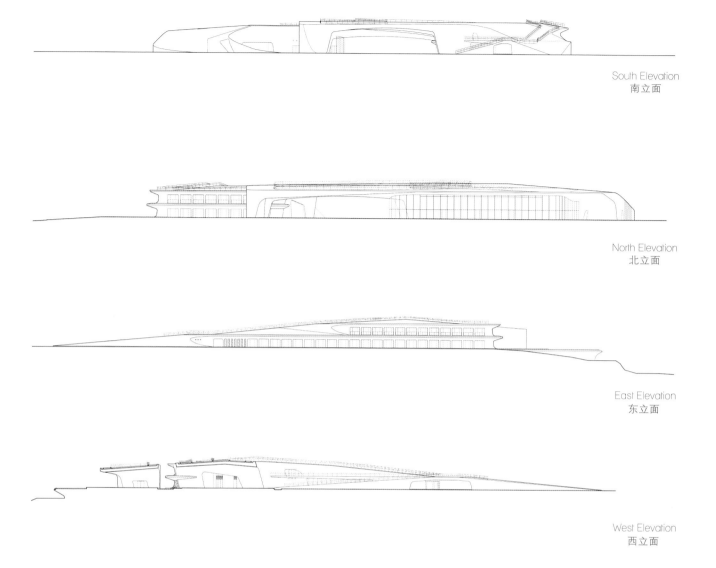

South Elevation
南立面

North Elevation
北立面

East Elevation
东立面

West Elevation
西立面

The site just touches the narrow inlet extending almost south – north at its northern tip, has a narrow opening facing the lake – view direction, and extends relatively deep inland along a road. Looking towards the lake, the lake surface looks like it is cutout in a V shape as mountain slopes close in from both sides.

One way to solve the first problem was to pursue a new relationship between the building and its surrounding landform. Since long ago, buildings have generally been built "on" landforms, but there have been cases in which they have been built within landforms, such as the Yao Tong settlements along the Yellow River. This half – architectural and half – landform project is conceptualized as a stage setting to bring out and amplify a hidden dimension of the scenery and environment of Sun Moon Lake, and at the same time create a new dialogue between the human being and nature that provides another new dimension to this area.

本案入口狭小，自南向北延伸。入口隧道中有一个狭窄的露天空地。这块空地面向日月潭，形似小道并逼近潭水，打造出极佳的观景视角。若细心地凝望，你会发现湖面因被两旁的山体相夹，呈现出 "V" 字形的切口。

本案试图以一种全新的方式演绎建筑与自然景观之间的关系。自古以来，建筑物大多建在自然景观之 "上"，作为主体被自然景观修饰。但是有些案例也证明了，建筑物也可建在自然景观之 "内"，作为客体衬托自然景观，比如黄河流域的窑洞住宅。"一半建筑，一半自然景观" 成为一种概念化设计构思。本案设计师试图通过一个隐藏的建筑，塑造全新的自然景观，衬托日月潭风景区之美。同时，本案设计师也开辟出一种全新的对话方式，厘清了建筑与自然景观之间的关系。

SHAPE ANALYSIS 造型分析

ECOLOGICAL SHAP

To avoid too much human intervention, the building located in the scenic spot sinks itself into natural environment in a low-key attitude. The higher side of the building gradually slopes down to the ground level at the other side then finally blend to natural environment harmoniously.

生态造型

为避免受到过多的人为干扰，本案选址于日月潭风景区内，建筑风格低调，流畅的建筑造型与大自然很好地融为一体。建筑从一侧高起，以逐渐下斜的方式过渡到另一侧，最后降落至草地并融入自然，呈现和谐之美。

SKIN ANALYSIS 表皮分析

ECOLOGICAL SKIN

In order to maintain nature, the roof of the building is covered by lawns in different colors that well hide the volume. The large rooftop lawns not only absorb sunlight radiation for interior and regulate micro climate, but also be a good publicity to the concept of environmental protection.

生态表皮

为保护日月潭的原生景观,本案屋顶被五颜六色的草坪所覆盖,将整个建筑体量隐藏了起来。屋顶上那大片大片的草坪,不仅为室内采光和微气候的调节作出了贡献,还宣传了环保理念。

CONGRESS CENTER IN KRAKOW

克拉科夫会议中心

Architect: Ingarden & Ewy Architects
Collaboration: Arata Isozaki & Associates
Client: Krakow City Council
Location: Krakow, Poland
Site Area: 16,000 m²
Photography: Ingarden, Ewy Architects

设计公司：Ingarden、Ewy 建筑事务所
合作设计公司：Arata Isozaki & Associates
客户：克拉科夫市议会
地点：波兰克拉科夫
占地面积：16,000 m²
摄影：Ingarden、Ewy 建筑事务所

STRUCTURE AND MATERIAL 结构与材料

STRUCTURE
Concrete Beam Structure

MATERIAL
Glass, Titanium Steel, Ceramic Panels, Granite, Lime Stone, Sand Stone Plates, and Historical Materials Used in Wawel Castle.

结构
混凝土梁柱结构

材料
玻璃、钛钢、陶瓷板、花岗岩、石灰、石板，以及瓦维尔城堡的典型古材。

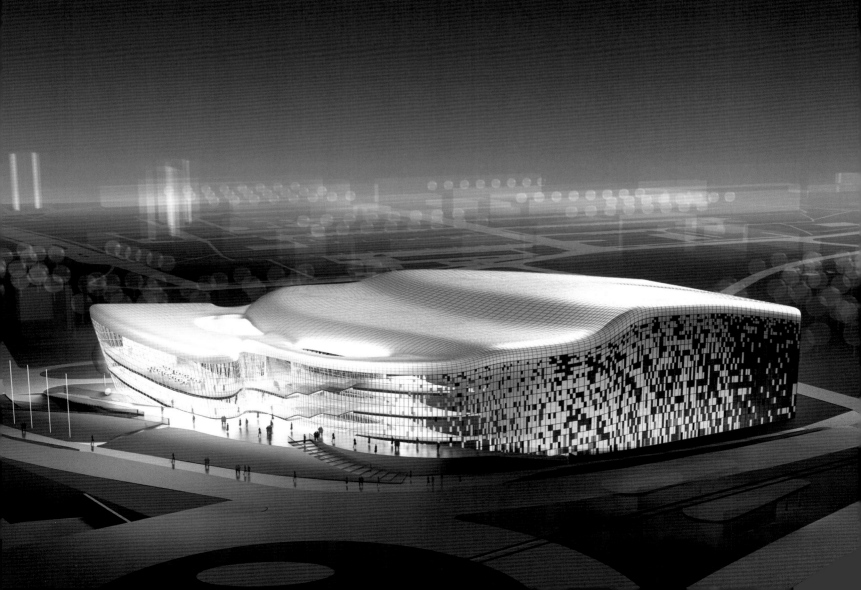

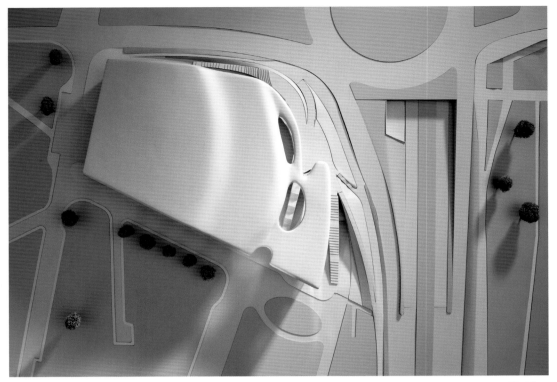

Site Plan
总平面图

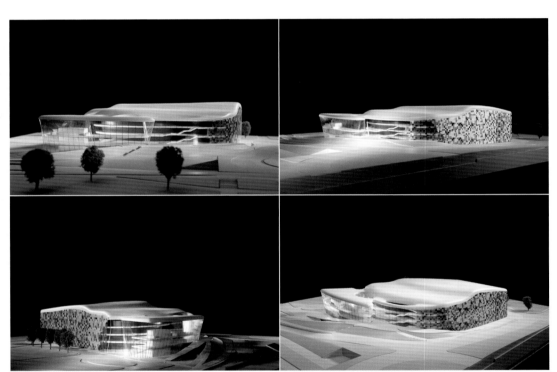

Congress Center will create a new architectural reference to a fragment of the Krakow city nearby Vistula river waterfront, adjusted to the prestigious historical and contemporary context of the Old City, Kazimierz and Podgorze districts. Functional and compositional building layout follows interior and exterior traffic directions scheme, determining view points to create extensive public spaces opened to the city`s main historical landmarks.

New Congress Center occupies a corner of the plot nearby Grunwaldzki Roundabout an important car, bus and tramroad junction.

会议中心将部分参照维斯瓦河畔的克拉科夫城市建筑风格，创建一个新建筑，以适应包括犹太广场和普格兹广场在内的古城区的新时代发展。建筑的功能和组成遵循其内外交通方位安排来布局的，朝向城市的主要历史建筑地标，创造出大量的公共空间。

新会议中心坐落于Grunwaldzki附近，位于一个汽车、公交和有轨电车交汇处。

Model
模型图

SHAPE ANALYSIS 造型分析

DIGITAL SHAPE

Building echo the contemporary science and technology, and use computer aided design the irregular shapes.

数字造型

建筑呼应了当代科技的发展，运用计算机辅助设计塑造出不规则的造型。

Sketch
概念草图

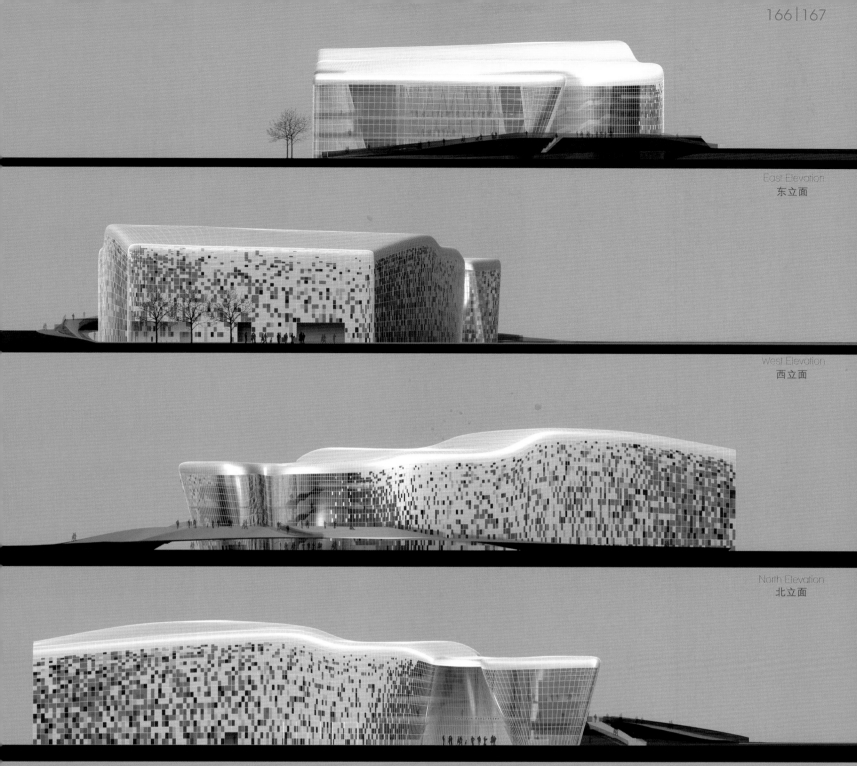

East Elevation
东立面

West Elevation
西立面

North Elevation
北立面

South Elevation
南立面

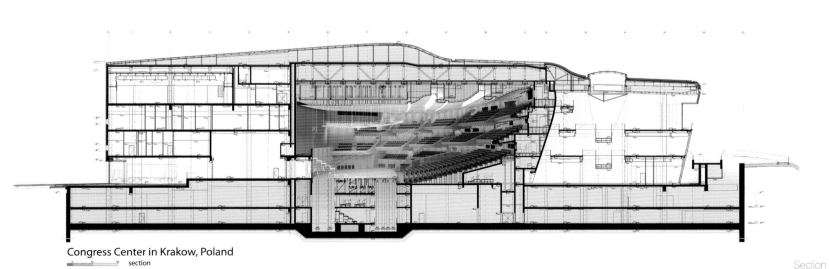

Congress Center in Krakow, Poland
section

Section
剖面图

SKIN ANALYSIS 表皮分析

LIGHT SKIN

According to the different environmental needs, building adopted a variety of skin materials, such as glass, titanium steel, ceramic plate, granite, limestone, slate, etc.

轻表皮

根据不同的环境需要,建筑采用了多种表皮材料,如玻璃、钛钢、陶瓷板、花岗岩、石灰、石板等。

Contemporary countenance of the building is formed by organic irregular shapes and differentiation of materials contrasted with spectacular foyer covered with glazing opened up to a great view of Krakow panorama. The solid dimension is diminished in direction of Vistula river due to optical reduction of a large building scale in context of green boulevards nearby. Glazed, three storey high foyer allows visitors 'to see and to be seen'.

Basic elevation materials, glass and titanium steel, are refilled with individually designed ceramic panels and granite, lime – stone, sand – stone plates, referring to typical historical materials used in Wawel Castle. Composition of differentiated materials juxtaposed with glass creates a colorful mosaique on elevations defining architectural expression of the building.

建筑的立面充满了现代气息,主要由不同的材质以不同的形状组合而成,与大门厅的玻璃釉面形成了鲜明的对比,能观赏到克拉科夫全景。由于降低了位于附近的林荫大道上的大规模建筑在视觉上的影响,朝向维斯瓦河的实体墙面面积也相应减少了。通过玻璃幕墙,三层楼高的大堂使到此的游客都成为了观光者和被观光者。

建筑的基本立面材料为玻璃和钛钢,与独立设计的陶瓷板、花岗岩、石灰、石板、以及瓦维尔城堡的典型古材相互搭配使用。玻璃配合以多种成分材料,共同创建了一个华美而富有生气的外立面,生动地传达了建筑的思想。

Congress center is around 37 000 m². The building contains 3 halls with a capacity of 1800, 600 and 300 seats, and conference rooms of 500 m², storages, technical supply and underground parking for 320 cars.

会议中心共占地 37 000 m²，包含三个分别可容纳 1800、600 和 300 个席位的大厅、一个占地 500 m² 的会议室、仓储室、技术供给室，以及一个可以容纳 320 辆车的地下停车场。

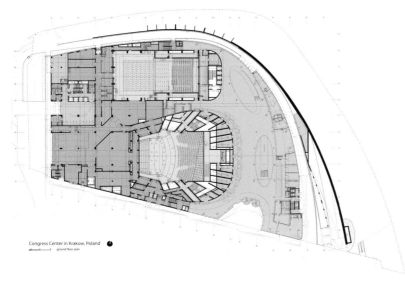

Ground Floor Plan
首层平面图

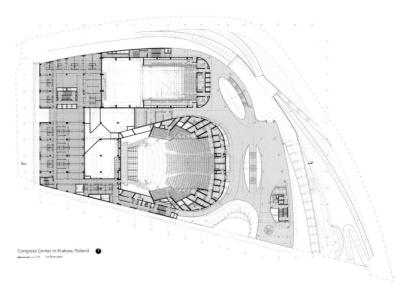

2nd Floor Plan
二层平面图

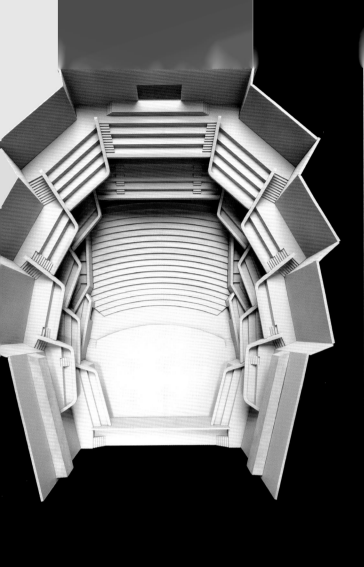

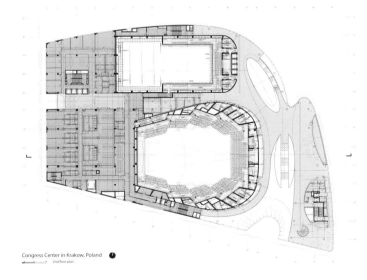

3rd Floor Plan
三层平面图

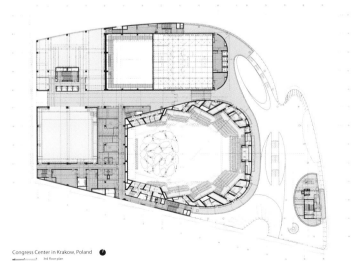

4th Floor Plan
四层平面图

PROSECUTOR'S OFFICE TBILISI

第比利斯检察官办公厅

Architect: Architects of Invention
Client: Ministry Of Justice Georgia
Location: Tbilisi, Georgia
Site Area: 3,000 m²
Photography: Nakanimamasakhlisi Photo Lab

设计公司：Invention 建筑事务所
客户：格鲁吉亚司法部
地点：格鲁吉亚第比利斯
占地面积：3 000 m²
摄影：Nakanimamasakhlisi Photo Lab

STRUCTURE AND MATERIAL 结构与材料

STRUCTURE
Framed Structure

MATERIAL
Glass, Steel

结构
框架结构

材料
玻璃、钢材

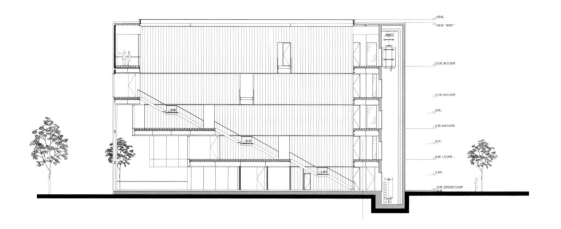
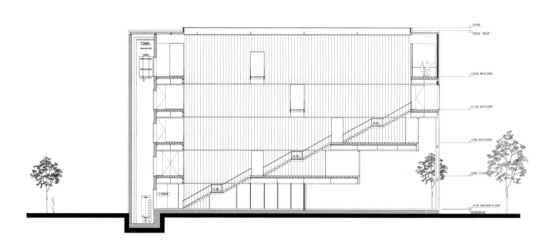
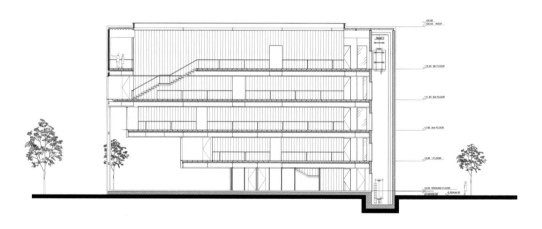
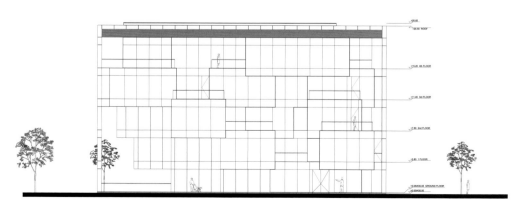

Section
剖面图

SHAPE ANALYSIS 造型分析

OPEN SHAPE

The front of the building looks formal and smooth, while on the back is an unexpected "floating" structure by irregular protrusions as a loggia or a balcony. The hollowed structure on one side of the building becomes the best transition between the blocks.

开放式造型

本案正立面规则而平整，但后立面却出人意料地采用了"悬浮式"不规则架构，而突出的地方就被用作凉廊或阳台。从侧面望去，建筑呈现出华丽的中空漂浮结构，这种结构能很好地让建筑自然和谐地过渡到街区。

The building consists of a stalwart, stark black frame that has luminous, delicate volumes suspended within it. 70% of the building is freestanding away from the sides and the ground creating an impression of 'floating' forms. The volumes make way for a passageway to one side of the building, creating a physical transparency of the building that is alluded to throughout the design. The suspended volumes are staggered so that they create terraces and balconies.

The interior intelligently uses glass to create a feeling of transparency and light, in contrast to the dark impenetrable framework. The ground floor serves as the main access area with security check in and lobby. The top floor is dedicated to the Prosecutor's office, a meeting room, canteen and veranda for use by staff. A sky park, or roof garden, is used for conferences and celebrations. In contrast to the openness of the front facing façade, the back of the building is smooth and functional with few windows.

The choice of materials and the form have been chosen to reflect the function of the building. The stark geometry of the black frame represents the uncompromising nature of the legal system. Inside the building, a clustered, mirrored volume endlessly throws reflections: "The semantic message here is that the centre is unknowable - purely because it is an endless reflection and ordering of the world around it", says Architects of Invention director Niko Japaridze. "The structure mirrors the brain; the glass exterior mimics the lobar matter which processes exterior realities, while the hypothalamus is the centre of the nervous system controlling temperature, hunger, thirst and fatigue".

本案采用一种稳定性极高的黑色框架。这种框架能打造出硬朗的整体造型，营造出精致悬浮的局部空间。本案 70% 的建筑呈自由分离的形态。空间之间彼此相距，远离地面，形成"悬浮"结构，而且中空之处能开辟一条通往建筑物另一边的小径。因为这种分离结构十分精巧，所以本案的总体特点被归纳为"通透"。所悬浮的空间排列错落有致，非常适合打造成露台和阳台。

本案室内巧妙地使用玻璃光面，它既能提高透明度和增强照明效果，又能中和框架造型那种暗沉硬朗的风格。室内第一层是业务大厅，设置了安全检查区和会客休息厅。顶层是专用的办公区域，有检察官办公室、会议室、餐厅和供员工使用的活动空间。而那些悬浮在空中的普通花园或屋顶花园，一般用作会议厅和派对场所。这种设计可比性特别强，背立面空旷开阔并有精致悬浮的不规则结构，而正立面却平滑规整得只有几个用于通风的窗口。

建筑材料和造型结构的选择取决于建筑物的功能与用途。黑色而沉稳的框架造型十分符合检察办公厅的形象，体现出不可抗拒的法律制度本质。而本案室内采用的大量玻璃镜面，具有引人深思的效果。Architects of Invention 总监 Niko Japaridze 认为："玻璃设计寓意法律是不可预知和纯洁无瑕的。法律不但引人思考，还让所有人遵守它的秩序。"此外，"建筑物的不规则造型映射人脑构造特点；会受外来环境影响的玻璃表面就如人体肺叶，而掌控温度、感知饥渴和感受疲倦的下丘脑就是神经系统的中心。"

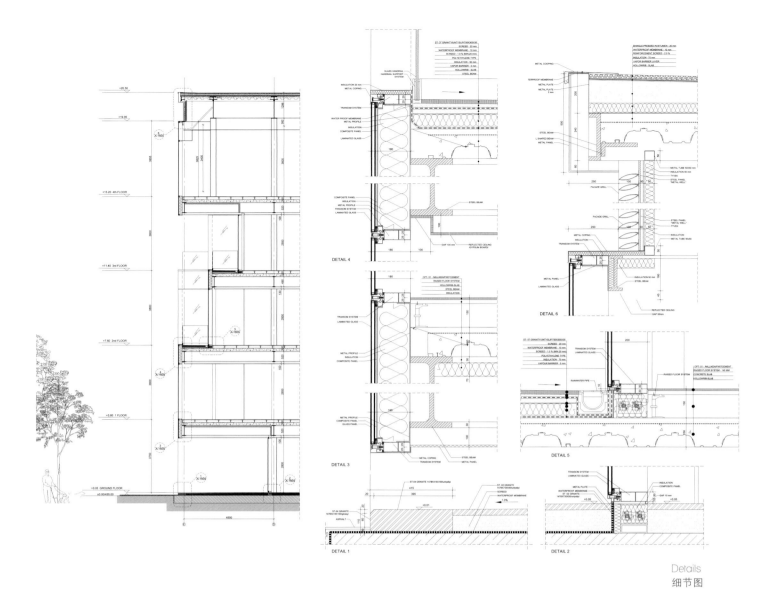

Details
细节图

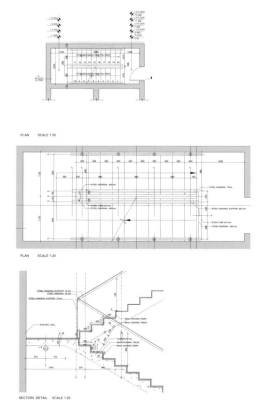
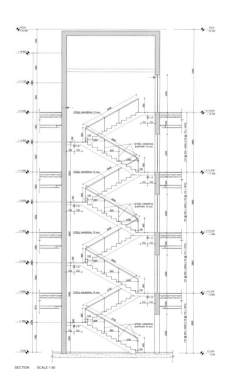

Details
细节图

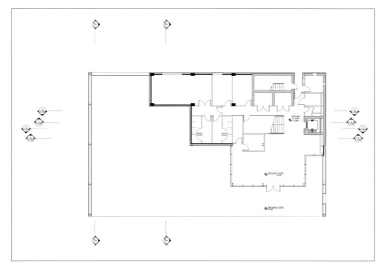
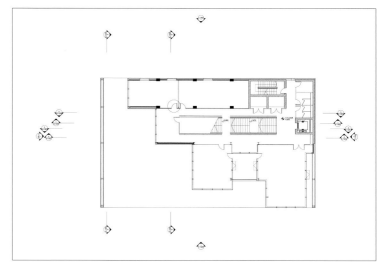
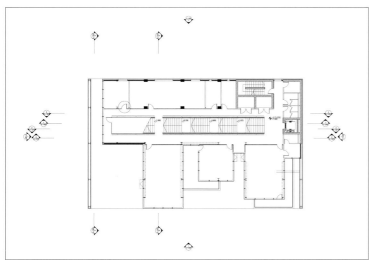
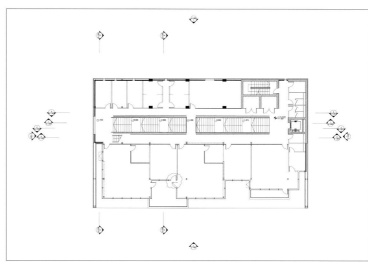
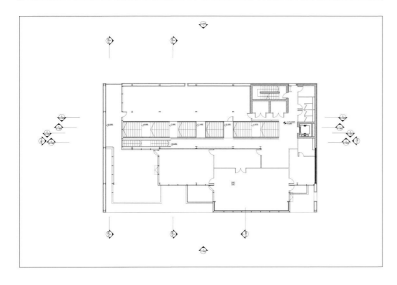
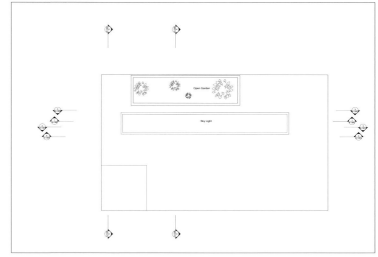

Floor Plan
楼层平面图

SKIN ANALYSIS 表皮分析

ENERGY SKIN
The interior intelligently uses glass to create a feeling of transparency and light, in contrast to the dark impenetrable framework.

节能表皮
因为玻璃能够营造透明高雅的室内环境，还能增强照明效果。本案室内非常巧妙地使用玻璃材质，让室内环境光洁明亮，恰好与外部结构暗沉硬朗的特色形成强烈的反差。

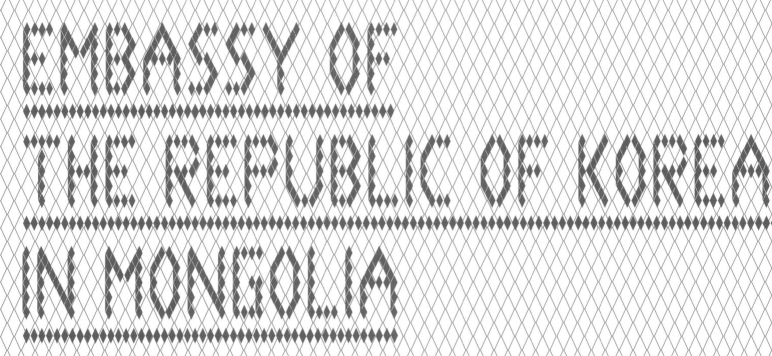

EMBASSY OF THE REPUBLIC OF KOREA IN MONGOLIA

韩国驻蒙古领事馆

Architect: Heerim Architects & Planners Co., Ltd
Client: Ministry of Foreign Affairs and Trade
Location: Ulaanbaatar, Mongolia
Site Area : 10,000 m²

设计公司：Heerim 建筑规划有限公司
客户：韩国外交通商部
地点：蒙古乌兰巴托
占地面积：10 000 m²

STRUCTURE AND MATERIAL 结构与材料

STRUCTURE
Reinforced Concrete Structure
MATERIAL
Concrete

结构
钢筋混凝土结构
材料
混凝土

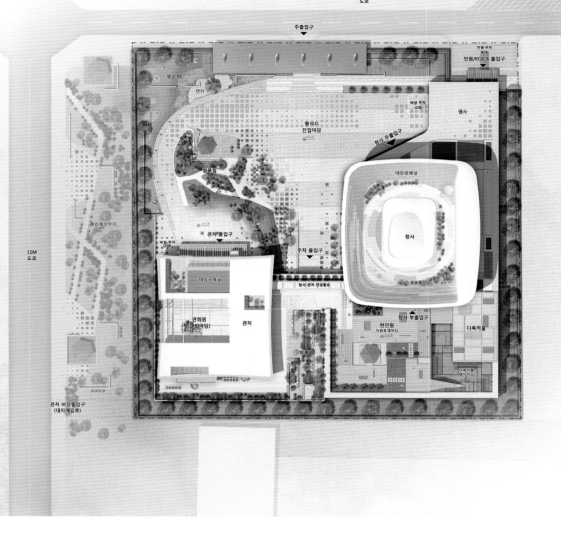

Site Plan
总平面图

The official residence has been divided into two defined zones which are visitor's zone and family zone, and the family zone has been planned as an eco-friendly residential space with a balcony and green areas.

Various traditional elements have been applied to the elevation design including lines of eaves, traditional Korean lantern "Cheongsachorong", patchworks and traditional pushup windows. Lines of eaves in traditional architecture have been adapted into the design of the roofline, and the traditional Korean lantern "Cheongsachorong" becomes a design motif to express a sense of warm welcome to the visitors. Also, the patchwork patterns have been used for the roof design to create a modern Korean inspired space. Furthermore, features of traditional push-up windows have allowed the use of outdoor space to be converted as a half-indoor space and flexible banquet space.

The government office reselects the form of "Korean Traditional ceramic" conveying the traditional culture that treats visitors with kindness, whereas the official residence has been inspired by the form of "Water container", called "YeonJak", used by the ancestors when writing a letter.

The façade of the government office creates an image of Korean landscape with traditional lotus flower patterns. Indoor connectivity of every facilities and High performance building skin are the main features considering the local climate, with long winter season with lows of -45 degrees Celsius. Moreover, the atrium placed in the center of the building is designed to save heating energy, while allowing natural ventilation and natural lighting.

官邸的建设包括接待区和家庭区两部分，其中，家庭区还设置有阳台和绿地，是一个生态友好型住宅空间。

设计融合了诸多韩国传统元素于其中，例如线性屋檐、传统韩式灯笼"Cheongsachorong"、拼缀物、以及韩国传统的上推式窗户。线性屋檐在传统建筑中被应用在屋顶的设计中，而传统韩式灯笼"Cheongsachorong"则作为一个全新的设计主题运用在本案中，传达出温暖的感觉，以吸引更多的访客。同样，拼缀物的造型也被融入到屋顶的设计当中了，以营造出一种能表现现代韩国灵魂的空间。此外，传统的上推式窗户将室外空间转化为一个半室内空间和一处灵活的宴会空间。

政府办公室重新选定韩国传统陶艺为形象来传达韩国的传统文化及其友善待客的理念，然而，官邸的设计灵感则来自于韩国古人在写信时使用的物品，具有"水容器"之称的"YeonJak"。

政府办公室的立面打造了一幅带传统莲花图案的韩国景观图。每一个设施的室内连通和高性能的建筑表皮都反映了当地独特的气候：冬季漫长，平均气温低于-45℃。另外，为了节省供热能源，同时为建筑引入自然通风和自然光照，设计师在建筑中心设置了中庭。

Volume & Function Analysis
体量和功能分析图

SHAPE ANALYSIS 造型分析

PICTOGRAPHIC SHAPE

Traditional lines of eaves, Korean lantern "Cheongsachorong", patchworks and pushup windows blended into the design of building, not only convey the traditional culture of Korean, but also endow the building with a warm hospitality.

象形造型

韩国传统的线性屋檐、韩式灯笼"Cheongsachorong"、拼缀物,以及上推式窗户都融入到建筑的设计当中,不仅展现了韩国的传统文化,更赋予建筑热情好客的温暖形象。

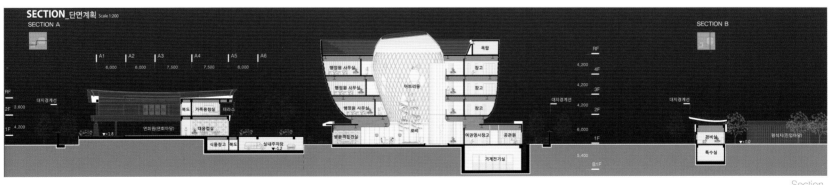

Section
剖面图

Elevation
立面图

SKIN ANALYSIS 表皮分析

ENERGY SKIN

The carved hollow pattern of skin gives the building traditional yet different kind of aesthetic. Such high performance skin keeps warm in interior, which is the best respond to the cold winter.

节能表皮

表皮上的镂空雕花图案赋予了建筑别样的传统美感,高性能的表皮设计提高了建筑的保暖效果,能充分应对寒冷的冬天。

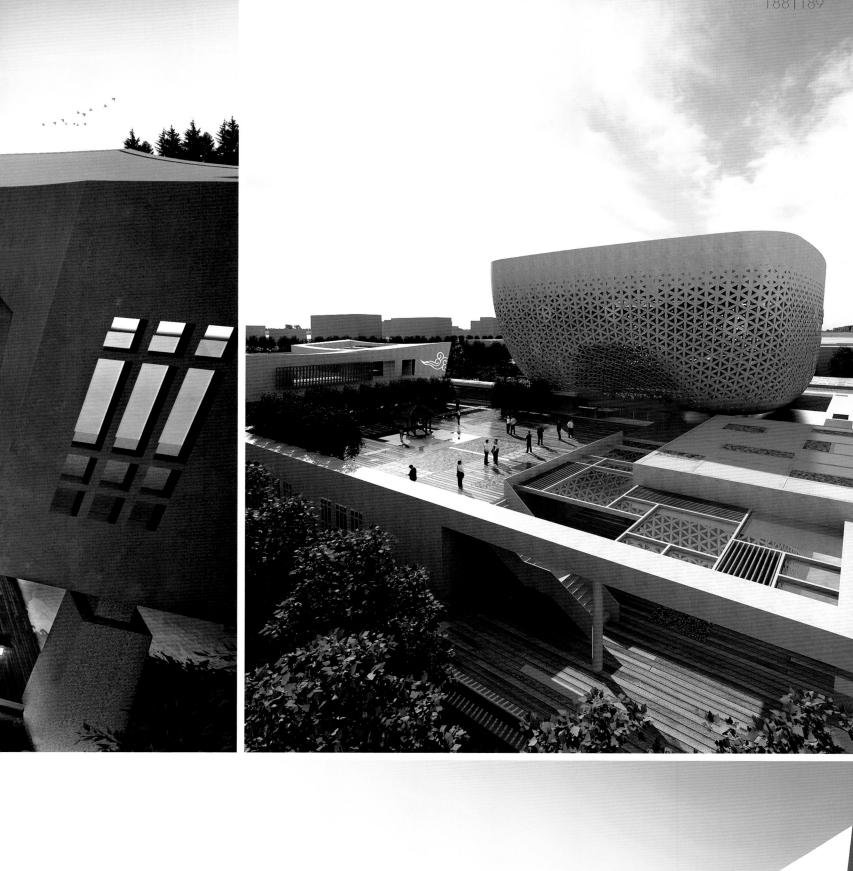

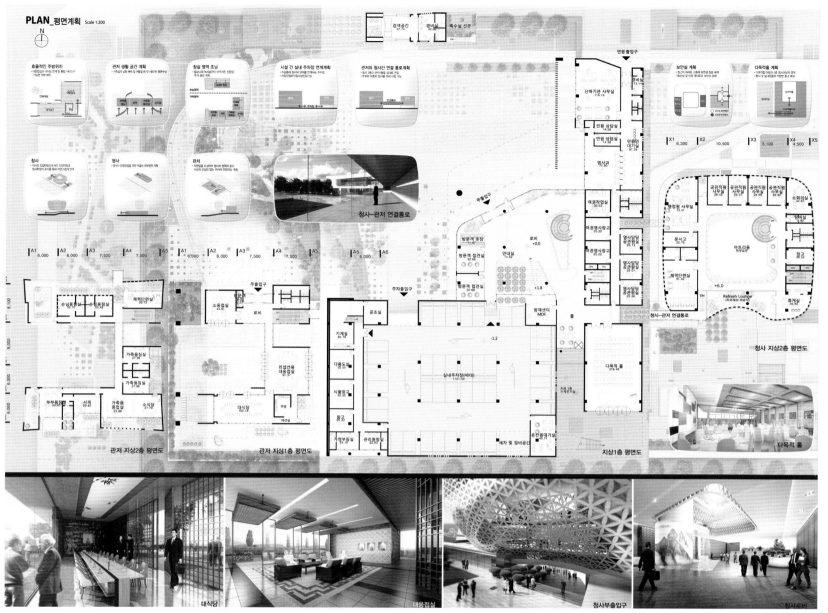

Floor Plan
楼层图

TIANJIN WEST RAILWAY STATION 天津西站

Architect: gmp·von Gerkan, Marg and Partners Architects
Client: Tianjin Ministry of Railway
Location: Tianjin, China
Built Area: 179,000 m²
Photography: Christian Gahl
Drawings: gmp·von Gerkan, Marg and Partners Architects

设计公司：冯·格康、玛格及合伙人建筑师事务所
客户：天津铁路局
地点：中国天津市
建筑面积：179 000 m²
摄影：克里斯汀娜·达尔
图纸：冯·格康、玛格及合伙人建筑师事务所

STRUCTURE AND MATERIAL 结构与材料

STRUCTURE
Steel Framed + Glass Façade

MATERIAL
Steel, Glass

结构
钢框架玻璃幕墙

材料
钢、玻璃

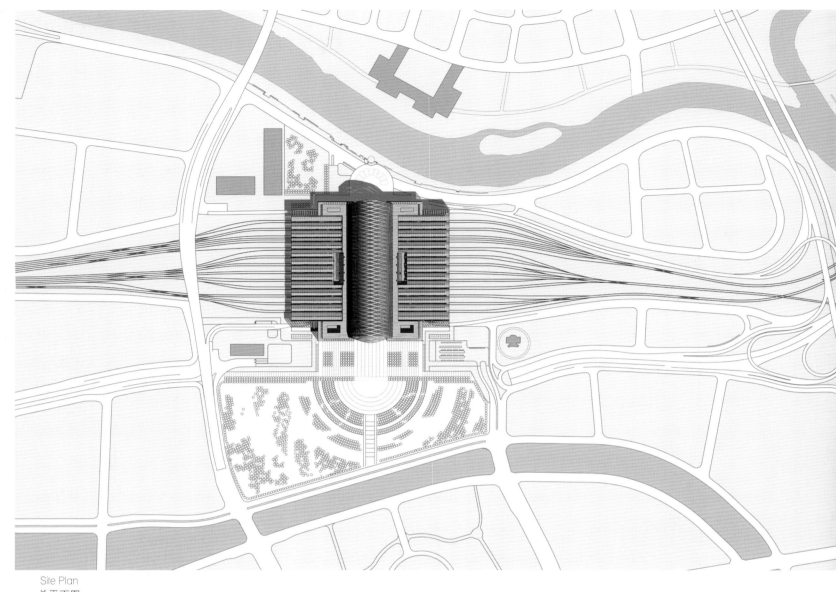

Site Plan
总平面图

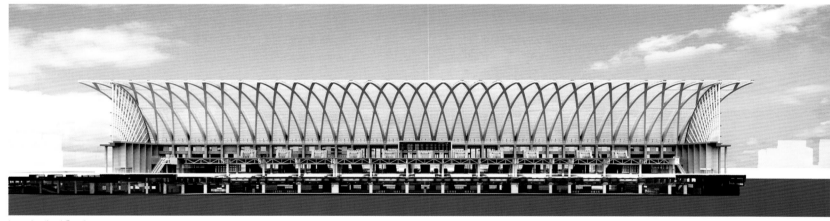

Longitudinal Section
纵剖面图

Cross Section
横截面图

The Tianjin West Railway Station, which is located about 130 kilometers south-west of Beijing, serves as a stop on the high-speed line between the Chinese capital and Shanghai, as well as connecting the various regional lines and linking these to the underground network. The local urban design function of the railway station is to connect a commercial area to the north with the old city center to the south, bridging tracks, a river and a road in this city of 12 million residents.

天津西站位于北京西南约 130 km 处，是京沪高铁线上的新枢纽，同时也是城际交通网与城市地铁网的交汇点。天津市拥有人口一千两百万，西站的落成在城市结构上实现了通过站轨、河运和公路将南部的老城区与北部的商务新区连为一体的目标。

SHAPE ANALYSIS 造型分析

PICTOGRAPHIC SHAPE

The architects have highlighted the bridge function between the city quarters with a 57 meter high and nearly 400 meter long barrel vault roof above the terminal concourse. Its curved roof is reminiscent of a large scale city gate and the long, stretched out concourse beneath of a classic place of transit. The portals of the eastern and western sides of the curved hall are symmetrically framed by arcades. To the south of the building a large and open station forecourt covers a wide area which gives credence to the importance and dimension of this railway station.

象形造型

本案的设计聚焦于建筑的连通作用，以一个高 57 m，跨度约 400 m 的拱形屋顶覆盖在主站房上方，象征着西站作为桥梁沟通着两个城区；巨大的卷拱传达了城市大门的意象，而其下方的长形大厅则以传统的车站形象呈现；主站房东西两侧的大门由两座拱形游廊对称围合着，与坐落于车站南侧的站前广场一起，构造了一个大尺度连续性的城市自由空间，体现了火车站的宏伟规模和文化内涵。

Passengers enter the new Tianjin West Railway Station through the main entrances on the north and south sides. Arched cantilevers above the entrances and tall window fronts convey an initial impression of the space passengers encounter in the concourse, which is flooded with daylight, providing a high quality atmosphere and clear orientation for travellers. The barrel vault roof conveys a dynamic impression, not least because its steel elements vary in width and depth from the bottom to the top, and are woven together. Escalators and lifts are available for passengers and visitors to descend to the platforms.

This technically and structurally sustainable railway station illustrates a contemporary interpretation of the cathedrals of traffic from the heydays of railway travel.

旅客可从位于南北两端的主入口进入新建成的天津西站。入口处高大的玻璃山墙立面顶部挑出的拱形悬梁是主站房设计理念的外在表征,刻画了宏大的空间,令人印象深刻:日光从屋顶透入大厅,营造了明亮舒适的室内氛围,使旅客可以清晰地找到各个空间。钢结构拱梁由下至上逐渐增粗增厚,穿插交织,勾勒出拱顶优雅而富有活力的整体形象。乘客可以从这里搭乘电梯或手扶电梯抵达下层的站台。

从技术设备到工程结构,天津西站均符合可持续发展要求,功能完备,设计者以极富当代精神的建筑表现手法,重新诠释了近代铁路运输鼎盛时期的经典车站建筑。

SKIN ANALYSIS 表皮分析

ENERGY SKIN

Daylight reaches the concourse through the diamond shaped steel and glass roof construction, and while the lower part is nearly transparent and admits a great deal of light, the upper part serves as protection against direct solar radiation. It reduce lighting energu comsumption effectively.

节能表皮

主站房的屋面为白色钢框架玻璃幕墙结构,日光通过其菱形网格射入室内;屋面底部的玻璃趋近透明,可过滤从顶部进来直射的阳光,更多地引入来自侧面的漫射光线,有效地降低照明能耗。

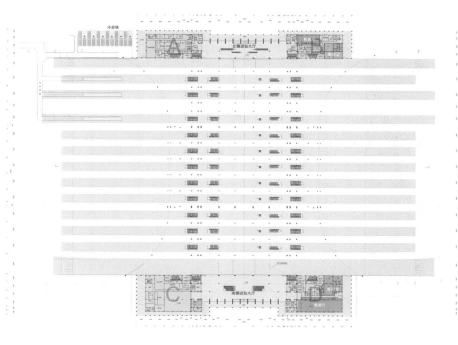

Platform Plan
站台平面图

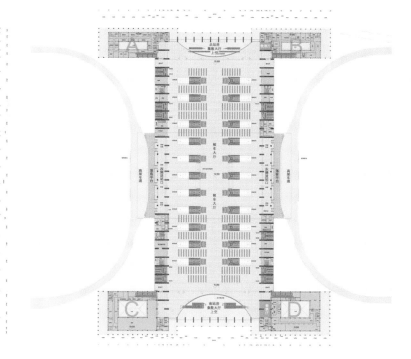

Waiting Hall Plan
候车大厅平面图

COMMERCIAL PLAZA

商业广场

CET BUDAPEST

布达佩斯 CET 建筑

Architect: ONL (Oosterhuis Lénárd)
Client: Porto Investment Hungary Kft.
Location: Budapest, Hungary
Build Area: 25,000 m²
Photography & Drawings: ONL (Oosterhuis Lénárd)

设计公司：ONL（Oosterhuis Lénárd）
客户：Porto Investment Hungary Kft.
地点：匈牙利布达佩斯
建筑面积：25 000 m²
摄影 & 图纸：ONL（Oosterhuis Lénárd）

STRUCTURE AND MATERIAL 结构与材料

STRUCTURE
Steel Framed Structure

MATERIAL
Steel, Glass

结构
钢架结构

材料
钢材、玻璃

Three of the 6 warehouses are now remaining, and the brief requests to keep at least 60% of the volume intact, while rightfully demanding to take away the first 20m of the 2 warehouses closest to the city to create a small square to improve the connection with the city south of the beautifully renovated Vásárcsarnok. Taking this into consideration, the design team proposes to develop the Közraktárak landmark complex in a smooth transition from old to new. The first two warehouse buildings will be carefully renovated while adjusting the size of the vertical windows as to open up the hermetic nature of the buildings to the Danube, to the interior gallery and reach out to the adjoining district with the successful Ráday utca nearby.

The river Danube fascinates in Budapest for its fast flow on its trajectory downward from the Schwarzwald to the Black Sea. While the Danube both separates and unites Buda and Pest, the CET aims at re-establishing visual contact at this point between the two sides of the river. Newly planned inviting terraces will visually open the once hermetic Közraktárak to the University and the Gellért Hotel. Hopefully a watertaxi system will be reintroduced to create direct connections for the people between the two sides as well. The body of the CET landmark building is developed along the flow of the Danube. Its architectural and urban expression evolves with the direction of the flow. The CET's origin stems from the side of the city centre, grows in size between the two parallel existing buildings of the Közraktárak and then culminates at the south side, the side of the National Theatre and the new Cultural Centre, in a striking landmark build¬ing representing the state-of-the-art in architectural design and building technology, its impact on the city will be not unlike the removed Elevator Building from the 19th Century from where the goods were distributed to the 6 warehouses which originally occupied the banks of the Danube.

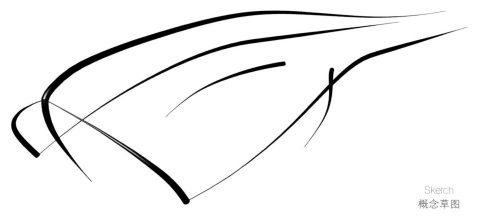

Sketch
概念草图

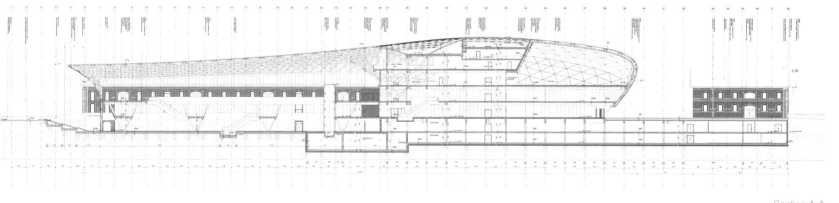

Section A-A
剖面图 A-A

在本案中，设计师保留了六间仓库中的三间，并按照客户的要求，保证了原址 60% 的完整性。放弃了离河岸最近的两间仓库，用以建造小广场，增强该区域与 Vásárcsarnok 以南的城市之间的连接。考虑到这一点，设计团队建议以一种更为温和的方式对 Közraktárak 地标建筑进行改造。在对前两座仓库进行翻新时，设计师调整了窗户的大小，使其能面向多瑙河边的建筑群及室内画廊，一改以前建筑物密不通风的环境，实现了该区域与相邻的 Ráday utca 的连通。

布达佩斯的多瑙河沿着河流从黑森林流向黑海，急流湍湍，震慑心魂，不仅起到分隔布达和佩斯的作用，还是两者间的纽带、联系着彼此。因此，CET 的设计目的是在河两岸重新建立一个视线连接点。在视觉上，最新设计的伸展阳台将面对着大学和 Gellért 酒店。此外，这里还有望重新引入出租快艇，为两岸的人们的交通提供便利。CET 标志性建筑的主体将沿着多瑙河岸边建设，其建筑形态和城市表现将随着河流的方向变化。CET 的开端始于市中心，与 Közraktárak 的现有建筑物呈平行排列，以南岸、国家大剧院和新文化中心处为终点，这是先进建筑设计和建筑技术的标志性建筑，其对城市的影响相当于以前的电梯楼对于原来那六间位于多瑙河两岸的仓库的影响。

SHAPE ANALYSIS 造型分析

DIGITAL SHAPE

CET is Central European Time. CET is also a synonym for a whale. The CET shape refers to the smooth and friendly streamlined body of a whale. Name and shape of the CET symbolizes its cultural potential and commercial pole position in one of the best preserved cities in the world.

数字造型

CET 是欧洲中部时间的缩写，也是鲸鱼的代名词。CET 建筑的造型模仿了鲸鱼流畅、友好的流线型身形。无论是名字还是造型，CET 建筑都象征着其是世界上保存最完好的古城之一，具有一定的文化潜力和商业价值。

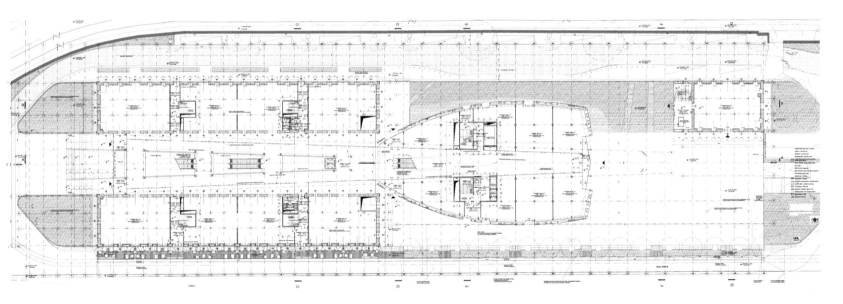

Ground Floor Plan
首层平面图

SKIN ANALYSIS 表皮分析

DIGITAL SKIN

Triangular panels are inlaided on transparent glazed façades that looked like covered with a layer of fish scales. This contributes to a change in the appearance of the original old buildings and gives the building a modern characteristic.

数字表皮

透明玻璃上镶嵌了三角形面板，让立面看起来如同覆盖了一层鱼鳞，一改原本老旧的建筑外观，赋予了建筑新时代的特征。

HANJIE WANDA SQUARE IN WUHAN

武汉汉街
万达广场

Architect: UNStudio
Client: Wuhan Wanda East Lake Real State Co., Ltd
Location: Wuhan, China
Built Area: 22,630 m²

设计公司：UNStudio
客户：武汉万达东湖置业有限公司
地点：中国武汉市
建筑面积：22 630 m²

STRUCTURE AND MATERIAL 结构与材料

STRUCTURE
Concrete Beam Structure

MATERIAL
Stainless Steel, Alabaster, Brushed Aluminum Panels

结构
混凝土梁柱结构

材料
雪花石膏、不锈钢、拉丝铝面板

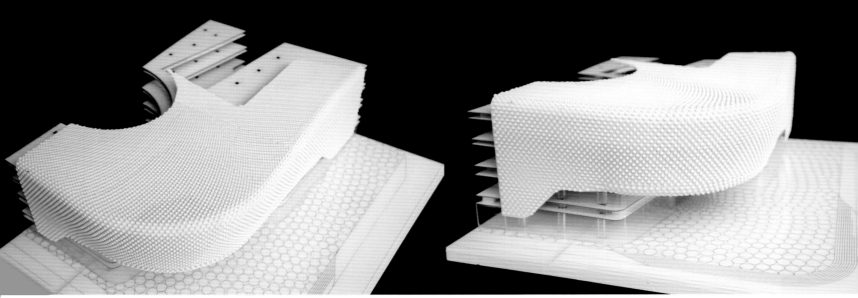

SHAPE ANALYSIS 造型分析

DIGITAL SHAPE

The façade design focuses on achieving a dynamic effect reflecting the handcrafted combination of two materials. The structure material, design, installation are completed through the parametric design.

数字造型

建筑的立面设计主要是为了营造出一种能反映两种材质的工艺性联系的动感效果。建筑的材料、设计和安装都是通过参数化设计得以实现的。

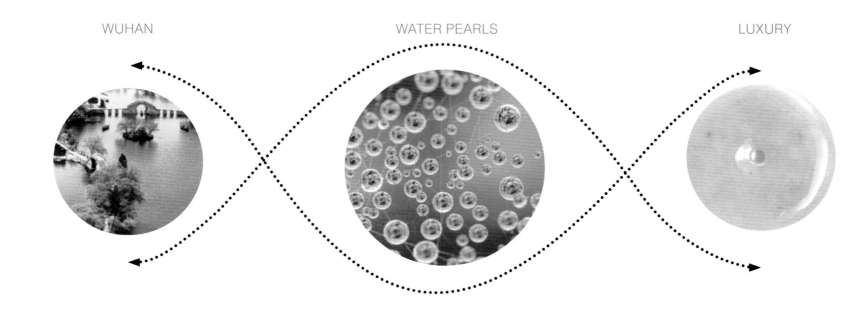

Material Concept
材料概念分析图

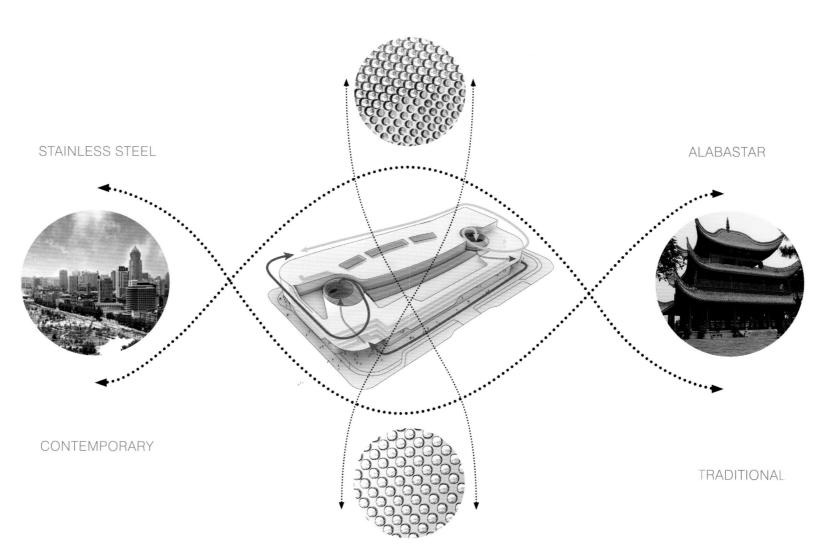

Connecting Synergy
材料联系分析图

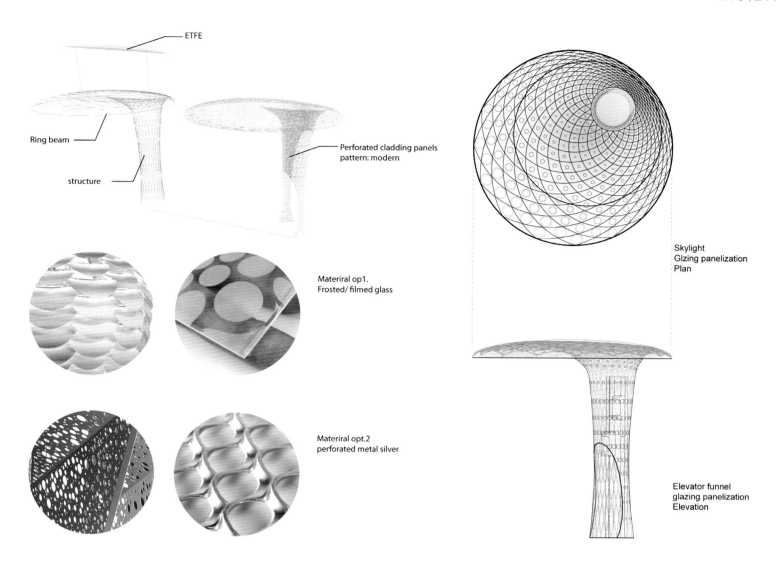

Hanjie Wanda Square is a new luxury shopping plaza located in the Wuhan Central Culture Centre. The multifunctional organisation of the master plan - which includes cultural and tourist facilities as well as commercial, office and residential components expected to be completed in 2013. UNStudio's design concept capitalises on the strategic location for the shopping plaza within the context of the master plan area. The concept of luxury is incorporated by means of ideas focussing on craftsmanship of noble, yet simple materials. UNStudio's approach considers the Hanjie Wanda Square as a contemporary classic, combining both contemporary and traditional design elements in one concept.

As water was utilised as a main organisational principle in the design for the Wuhan Central Cultural Centre, the theme "synergy of flows" is used as a starting point for the organisation of the buildings. The concept of synergy of flows is key to all of the design components; the fluid articulation of the building envelope, the programming of the façade lighting and its content design, the pattern and space articulation of the public landscape surrounding the building and the interior pattern language which guides customers from the central atria to the upper levels and throughout the building via linking corridors.

The interior concept is developed around the North and South atria, creating two different, yet integrated atmospheres. The atria become the centre of the dynamic duality of the two Hanjie Wanda Square identities: Contemporary and Traditional. Variations in geometry, materials and details define these differing characters. The North atrium is characterized by warm golden and bronze materials reflecting a cultural, traditional identity. In the South atrium silver and grey nuances with reflective textures reflect the city identity and its urban rhythm. Both atria are crowned by skylights with a funnel structure which connect the roof and the ground floor, in addition to integrating the panorama lifts. While the atria have strong and distinct identities, the corridors act as connectors between the two yet maintain their own character.

SKIN ANALYSIS 表皮分析

MULTIMEDIA SKIN

The building's skin can be controlled by computer to create light displays which generate glowing circular spots whilst simultaneously creating diffuse illumination.

多媒体表皮

建筑物的表皮可由计算机进行控制，形成光显示屏，循环播报广告，同时向四周漫射。

汉街万达广场位于武汉中央文化中心，是一个集文化旅游、商业办公和住宅于一体的多功能豪华购物广场，将于2013年竣工。本案的设计理念充分利用了项目所在的区域环境优势，用高档的工艺、简单的材料打造出奢华的感觉。设计师认为，汉街万达广场作为当代经典，应当将现代与传统的设计元素完美结合。

武汉中央文化中心的设计以水为主要元素，并以"流动中的协和"这个主题为建筑结构设计的出发点，和所有设计部件的关键所在。连贯的围护结构、外立面显示屏及其内容的设计、建筑周围的公共景观的造景及其空间的规划，以及建筑内部从中庭到楼上、到所有走廊上的指示标语，都体现了设计的主题。

内部的设计理念则围绕南、北前庭推进，创造出两个截然不同、但却浑然一体的氛围。前庭以其几何形状、材料和细节的变化，体现了汉街万达广场传统与现代两个动态特征。北中庭以温暖的金色和青铜色材料为主，体现了建筑的文化和传统的气息。南中庭则通过在一些细微之处采用带有反光效果的银色和灰色材料，体现了都市的形象和节奏。两个中庭都配备了全景升降机，中庭处还设置了漏斗形的天窗，将屋顶和地面连在了一起。而由于两个中庭都各具鲜明的特征，因此设计师采用廊道作为它们之间连接，保留了各自的特色。

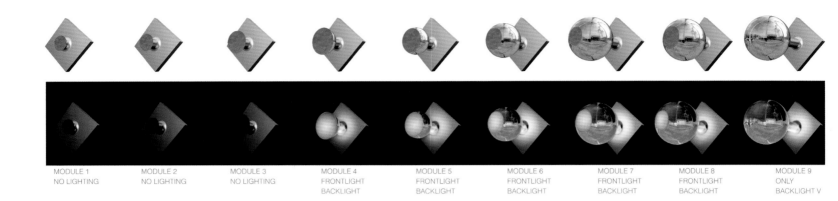

MODULE 1	MODULE 2	MODULE 3	MODULE 4	MODULE 5	MODULE 6	MODULE 7	MODULE 8	MODULE 9
NO LIGHTING	NO LIGHTING	NO LIGHTING	FRONTLIGHT BACKLIGHT	FRONTLIGHT BACKLIGHT	FRONTLIGHT BACKLIGHT	FRONTLIGHT BACKLIGHT	FRONTLIGHT BACKLIGHT	ONLY BACKLIGHT V

FACADE MODULES ARE CATEGORIZED AS "CHARACTERS" OF LIGHT. EACH OF THE LARGER MODULES BECOMES A UNIQUE LIGHT OBJECT, A FACADE ELEMENT AS WELL AS A SOPHISTICATED AND STYLISH PRODUCT IN ITSELF.

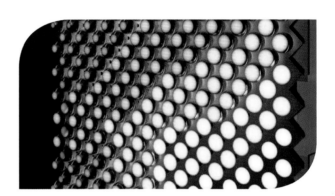

GRADIENT STUDIES

Façade Lighting
立面灯

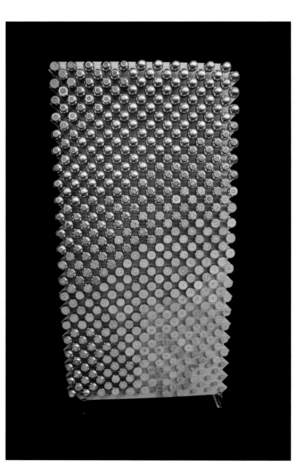
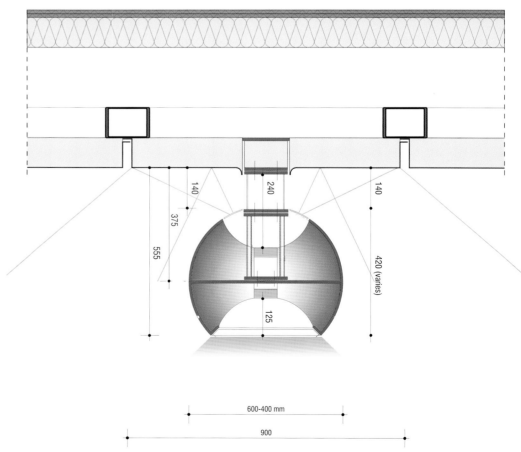

Façade Detail
立面细节图

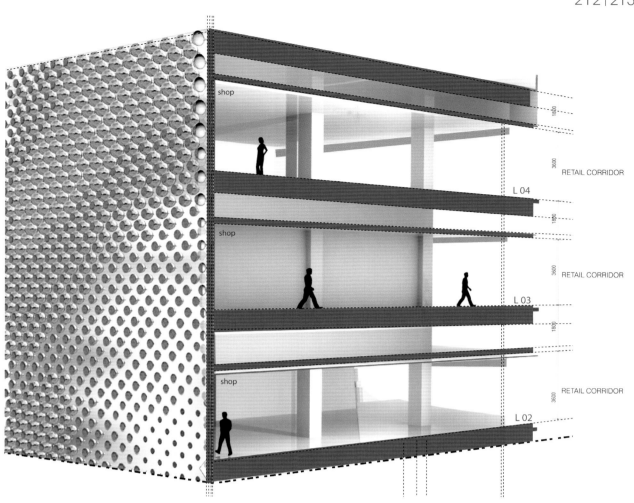

Longitudinal Section
纵剖面图

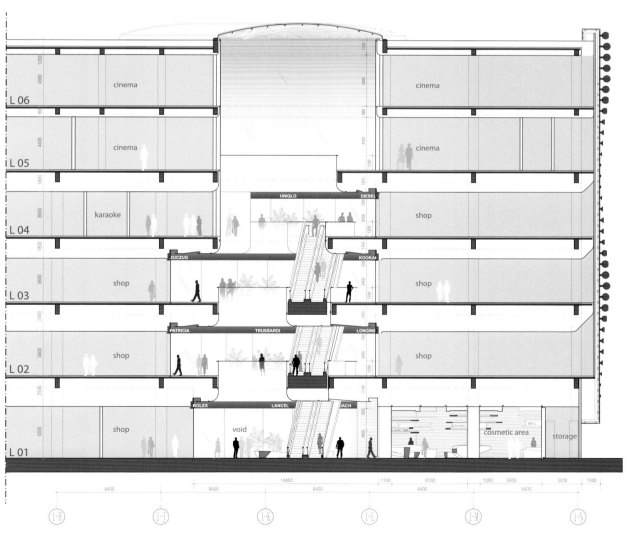

Longitudinal Section
纵剖面图

J.CUBE
裕冰坊溜冰场

Architect: RSP Architects Planners & Engineers (Pte) Ltd
Façade and Interior Concept Consultant: Benoy
Client: CapitaMall Trust
Location: Singapore
Site Area: 8,533.70 m²
Photography: Image Courtesy of CapitaMalls Asia Ltd,
RSP Architects Planners & Engineers (Pte) Ltd

设计公司：雅思柏设计事务所
立面设计和室内设计顾问：贝诺建筑事务所
客户：嘉德商用产业信托
地点：新加坡
占地面积：8 533.70 m²
图片提供：嘉德商用产业有限公司、雅思柏设计事务所

STRUCTURE AND MATERIAL 结构与材料

STRUCTURE:
Masonry Construction
MATERIAL:
PC panel, C-shaped Aluminium, Glass

结构
砌筑结构
材料
塑料板、弧形铝合金、玻璃

The name J Cube, already reflects every bit of its role. It boasts Singapore's first Olympic-sized ice-skating rink, a multiplex cinema with the first IMAX theatre in the suburbs and a roller-blading rooftop park.

One fundamental starting point for the design development of the façade arose from the Owner's decision for the building to appear more as an opaque box than a see-through glass box.

We opted to realize this opaque box via masonry construction. In order to depict the metaphor of fissure lines of ice, each pc panel was patterned with a composition of groove lines subtractively-formed at the point of casting each panel. The opaque panels were then coated with textured coating system of a light aqua-blue hue, to further the metaphor of ice. Plastering was not required, because the panels already had a good finish quality right after casting.

The diagonal patterning of the pc panels gave us the subtle fissuring we wanted, but we also wanted to further augment the "ice fissure" theme by having stronger diagonal lines. These stronger diagonals were formed by utilizing C-shaped aluminium extrusions. We decided to have the C-shaped aluminium extrusions surface mounted onto the pc panels (with a small gap to allow the free vertical flow of rainwater), in order to preserve the continuous water-tight skin afforded by the pc panels.

We were able to achieve a very good ETTV (of less than 40W/m^{2}K) even though we utilized single-glazed tinted glass, without double-glazing and without low-E coating. This was because of our very low Window-to-Wall Ratio – there were very little glazed areas relative to the total façade surface area, so even without the use of high-performance glazing we were able to achieve a building envelope of very low heat gain.

本案名为裕冰坊，从名字上就可以反映出其用途。它是新加坡第一个拥有奥林匹克规模的溜冰场、首家带3D影院的郊外多元剧院，同时也是一个屋顶旱冰场。

建筑的外观设计以业主的愿望为根本出发点，就是使建筑看起来就像一个不透明的盒子，而不是一个可透视的玻璃盒子。

为了实现"盒子"的不透明性，设计师选择了砌筑结构。为了表达出冰块中的裂纹感，设计师令每块塑料板在制造时就先预制好凹槽，建造时则根据凹槽线组成图案。这些不透明的面板被涂上了一层水蓝色质感的涂层，以进一步描绘出冰块的感觉。铸造后的面板具有了良好的光洁感，免去了抹灰的必要。

塑料板的对角线给设计师呈现了一种他们想要的细微裂纹感，但是设计师希望通过更强势的对角线来进一步展现冰块裂隙的主题。这些强势的对角线将采用弧形铝合金压材构成。设计师将这些弧形铝合金材料覆盖到塑料板上作为表皮，并在表皮上留下小缝隙让雨水垂直流下，以保护塑料板的持续水密性表皮。

建筑虽然采用了单层茶色玻璃，而不是双层玻璃，也没有增加低辐射镀层，但信号照样良好，可以清晰地收看东森电视台（导热系数少于40W/m^{2}K）。由于建筑的窗墙比较低，其外表采用的玻璃区域也相对较少，因此，即使没有使用高性能玻璃，设计师也能创造出一个低热量的建筑结构。

SHAPE ANALYSIS 造型分析

PICTOGRAPHIC SHAPE
Taking the shape of an ice but hidden its transparency, it's added an extra solid texture of a diamond.

象形造型
本案的造型灵感来源于冰块，却没有了冰块的透明感，多了一分菱角分明的坚实质感。

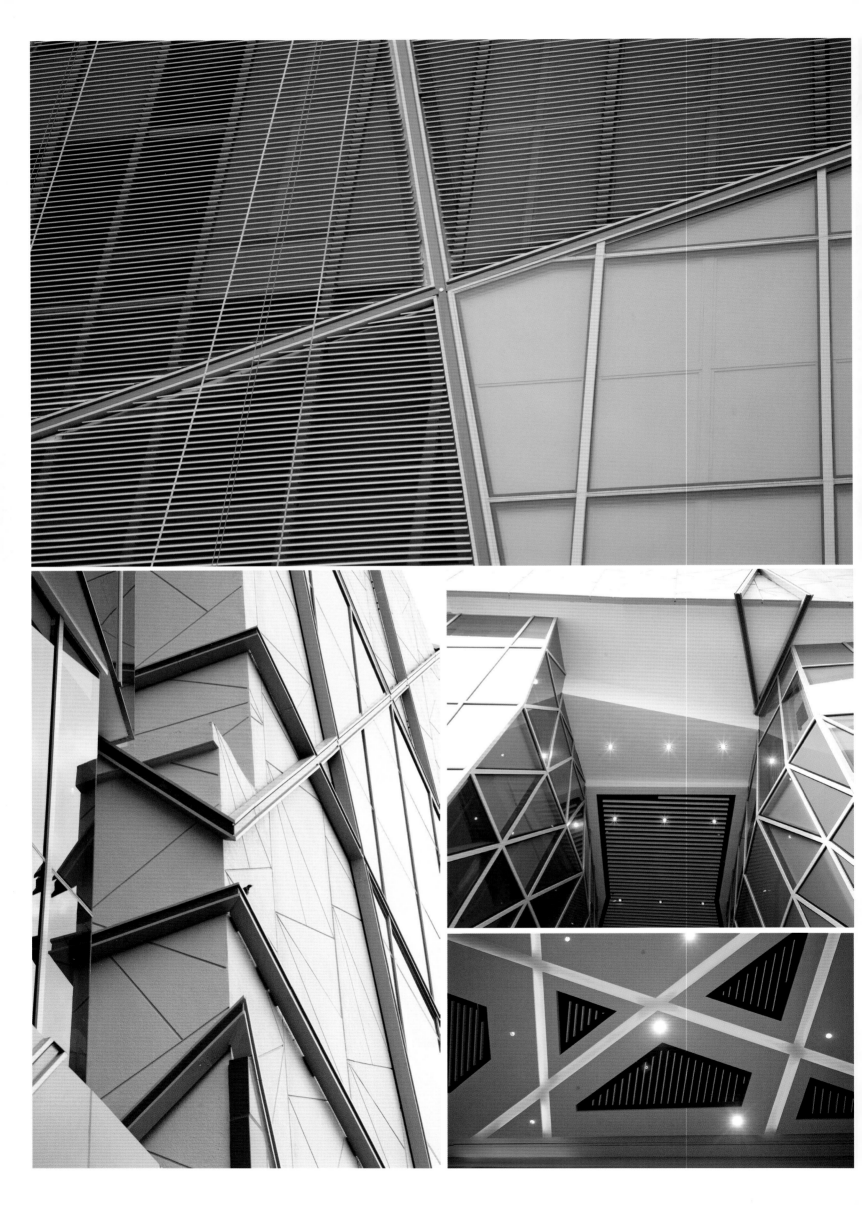

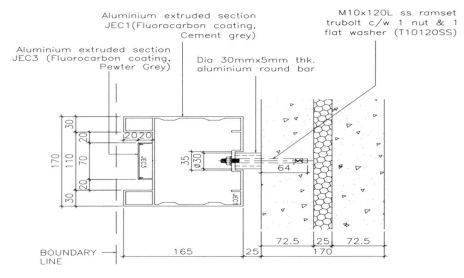

C-channel Alum Extrusion Detail
C 槽铝板挤压细节图

SKIN ANALYSIS 表皮分析

LIGHT SKIN

With the PC panels and glass make-up, the façade is interwoven in transparency and shade, solid but light.

轻表皮

立面由塑料板和玻璃组成，透视感和遮蔽性相互共存，坚固又不乏轻盈感。

MEDIACITE LIEGE BELGIUM

比利时梅迪思黛购物中心

Architect: Ron Arad Architects
Client: Wilhelm & Co
Location: Liege, Belgium
Built Area: 9,750 m²
Photography: Peter Cook, Marc Detiffe

设计公司：Ron Arad 建筑事务所
客户：Wilhelm & Co
地点：比利时列日市
建筑面积：9 750 m²
摄影：Peter Cook、Marc Detiffe

STRUCTURE AND MATERIAL 结构与材料

STRUCTURE
Steel Framed Structure

MATERIAL
ETFE Cladding, Aluminium, Glass, Steel

结构
钢架结构

材料
塑料覆层、铝板、玻璃、钢材

造型分析

数字造型

Aerial View Indicating Zones
各区域鸟瞰图

SHAPE ANALYSIS 造型分析

DIGITAL SHAPE

The exaggerated semi-elliptical components neatly arranged as fish scale is the main ridge of the entire building to support structure and layout dramatically.

数字造型

夸张的半椭圆部件如鱼鳞般整齐排列着,作为主脊支撑着整个建筑的结构和布局,颇具戏剧性。

Ron Arad Architects were invited by Wilhelm & Co to design a new shopping mall within the 40,000 "Mediacite" development. Situated in Liege, once the world's foremost centre of steel production and since in economic decline, the building stands out as a symbol of the city's revitalisation and strives to spearhead the city's regeneration.

The 350 m long mall weaves through the fabric of the refurbished old market centre at one end, through the new two storey building, connecting to the new Belgian national television centre at the other. The design of the roof unites these elements with a complex network of steel roof ribs that undulate through the mall. The lattice of steel sculpts the volume of the mall beneath, varying both in height and structural depth to form a variety of differing experiences. The steel ribs overhead, mirrored in the floor pattern, draw a sinuous pathway pulling you towards and through each of the zones, revealing diverse vistas that surprise along the way. As the structure exits in the volume of the main building (at the 2 Piazzas and at the link between the old market and new mall), the steel ribs wrap downwards, merging into façade to enclose the building's envelope.

Ron Arad 建筑事务所受到 Wilhelm & Co 的邀请，在占地 40 000 m² 的梅迪思黛开发区中设计一座大型购物中心。本案坐落在曾经是世界上最先进钢铁生产中心的比利时列日市，将作为城市复兴的标志，在经历了经济萧条后，带领城市走向重生。

购物中心总长 350 m，它的一端横穿被翻新的旧市场中心，另一端贯穿一栋新建的两层高大楼，与新比利时国家电视中心相连。通过复合网状的钢屋顶框架的使用，设计师将建筑顶部的设计与这些贯穿购物商场的通道相互连接起来。格架状的钢筋勾勒出购物中心下层的体量，呈现出变化多端的高度和结构深度，产生层出不穷的视觉体验效果。首层的地板与架空的钢筋框架设计相若，打造出一条蜿蜒曲折的走道，引领人们穿过各个区域，并于沿途欣赏购物中心内的不断变化的场景。作为主建筑的结构（分布在两个拱顶长廊中，并作为旧市场和新购物中心的连接），钢筋框架向下环绕，与外观形同一体，将建筑包围了起来。

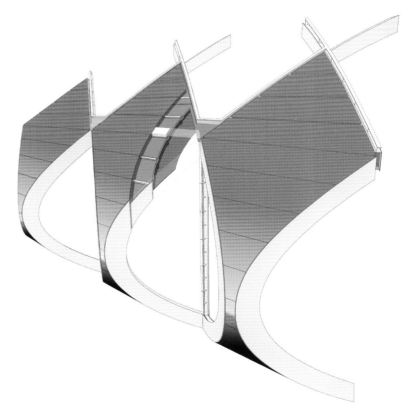

Link Cutaway
链接的切角部件

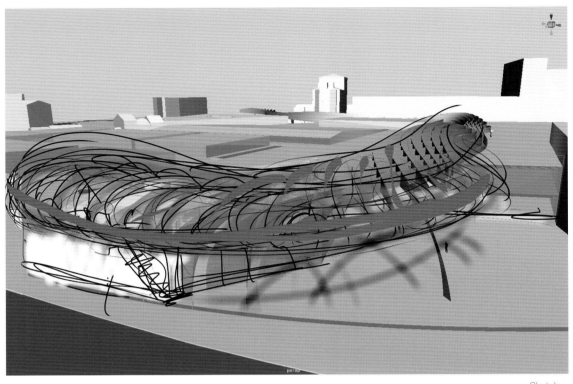

The structure is entirely free-spanning along its length and width, with 200 mm wide steel ribs that vary in depth from 1.2 m to 300 mm, weaving through each other in a deformed grid-like network.

The project was completed to a highly accelerated programme, with a duration of only 34 months from inception to completion. The total built area for the Mall is 9,750 m². With a construction budget of € 18.75 m the project was delivered for € 1,920/m². Construction began in April 2007 and the building was inaugurated in October 2009.

在建筑的纵深和总长设置中，设计师采用宽度为 200 mm，深度从 1.2 m 到 300 mm 不等的钢筋框架，相互交织成变形的格架网状，代替了跨度结构的使用。

本案的建设过程非常短，从开始到结束只持续了 34 个月。购物中心的总建筑面积为 9 750 m²，建设预算达 18 750 000 欧元，每平方米售价 1 920 欧元。2007 年 4 月动工，2009 年 10 月举行开业典礼。

Sketch
概念草图

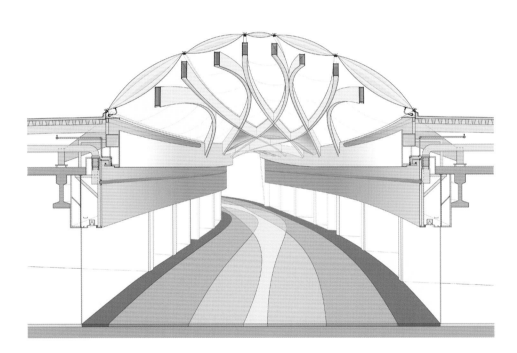

Perspective Section
透视剖面图

SKIN ANALYSIS 表皮分析

TRANSPARENT SKIN

To minimise loadings the complex 3 dimensional form is clad in transparent lightweight ETFE pillows – pneumatic Teflon cushions which allow light to penetrate the roof while moulding themselves to the irregular structure. As the roof gradually transforms into façade the ETFE cladding merges into curved aluminium rain-screen panels and glass.

透明表皮

为了将承重减到最小，建筑的立体形态上覆盖有透明的轻质塑料垫——以充气聚四氟乙烯垫制成，在保证光线能穿过屋顶的同时又能作为建筑不规则结构的组成部分。随着屋顶逐渐过渡到立面，塑料覆层也会和弯曲的铝质滤雨面板和玻璃融合到一起。

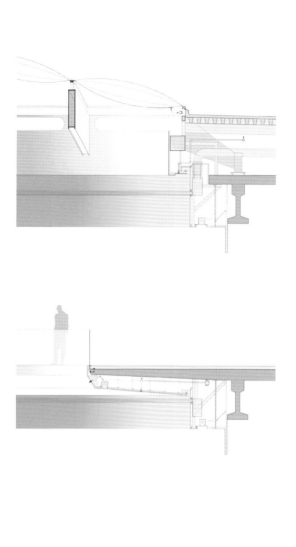

Detail Section
节点剖面图

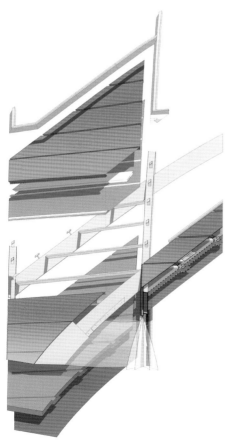

Link Detail
连接节点图

Sketch
概念草图

THAIHOT CITY PLAZA MALL
泰禾城市广场

Architect: Spark
Client: Thaihot Group
Location: Fuzhou, China
Built Area: 294,252 m³
Renderings: Bofan, Zhangpeng

设计公司：Spark
客户：泰禾集团
地点：中国福州
建筑面积：48 107 m²
效果图：Bofan、Zhangpeng

STRUCTURE AND MATERIAL 结构与材料

STRUCTURE
Reinforced Concrete Structure

MATERIAL
Aluminium Panel, Glass

结构
钢筋混凝土结构

材料
铝板、玻璃

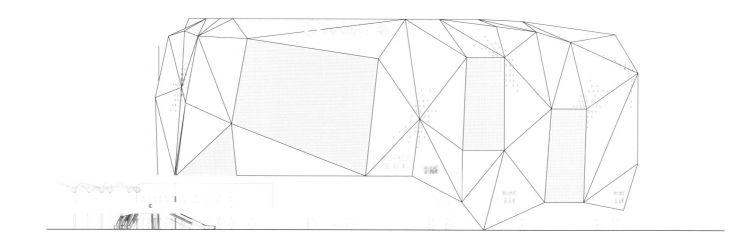

Elevation
立面图

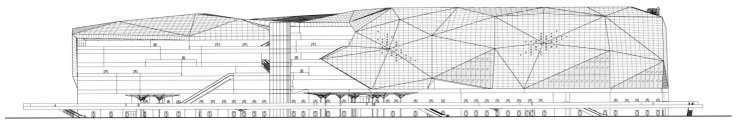

Elevation
立面图

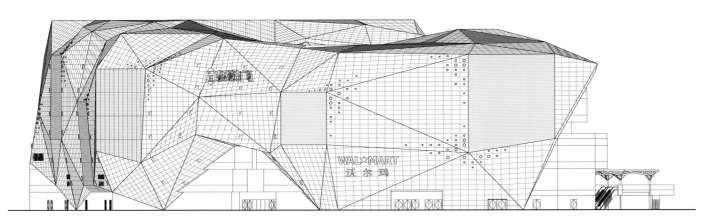

Elevation
立面图

Spark is working on the interior and façade design of Thaihot Mall Fuzhou. The multi-faceted façade has different functions, in plan and elevation, moderating the edges of the building, in plan increases sightlines into the pedestrian street drawing people in, unlike linear streets that fail to hold and capture the pedestrians' interest. The undulating tenant façade mimics the ebb and flow of pedestrian traffic creating a dynamic shopping experience.

Color shifting aluminium panels combined with signage and advertisements façade create an exterior appearance that is constantly changing. At night the perforated aluminium panels allow light to pass through to create a "starry night" effect.

设计师 Spark 负责本案的室内和立面设计。无论是从平面还是立面的角度来说，外立面所采用的多面体造型都具有多种功能，其不规则的造型能缓和建筑边缘，具有增强视觉效果的作用，能吸引街道行人的注意，而不像笔直的街道那般单调，让人兴味索然。往来不断的行人通道和川流不息的交通状况投射到此起彼伏的外立面表皮里，为人们带来了一种活力奔放的购物体验。

挂着广场招牌和贴着宣传广告的渐变色铝板使得建筑物外观不断变化。漆黑的夜空下，多孔铝板具有反射光线的作用，营造出"不夜城"的效果。

SHAPE ANALYSIS 造型分析

OPEN SHAPE

For it is a shopping plaza, the building is enclosed by concave-convex polyhedrons to deliver emphasis on openness. Gaps at the bottom are inlaid with transparent glass façade and the open atrium square links indoor to outdoor natural environment to offer a pleasant shopping experience for customers.

开放式造型

因为本案是一个购物广场，所以设计师打造出凹凸有致的多面体造型，以突出购物广场的开放性。底层的几个缺口镶嵌着透明的玻璃立面，空旷的中庭广场加强室内外与自然环境之间的亲密接触，让顾客体验到愉快的购物经历。

Ground Floor Plan
首层平面图

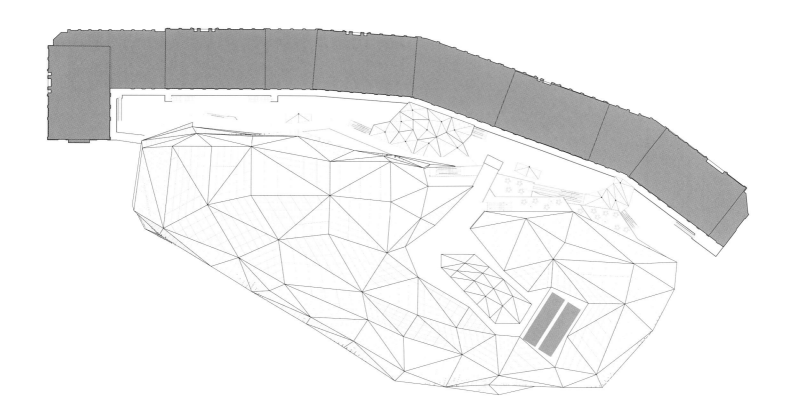

Aerial Elevation
顶部立面图

SKIN ANALYSIS 表皮分析

DIGITAL SKIN

The façade made up by polyhedrons forms different angles at different views, showing a play of light and shadow under sunlight or lighting that offering a colorful and changing effect.

数字表皮

凹凸不平的多面体立面朝向不同的角度，形成不一样的景观，在阳光或灯光的照射下上演光与影的游戏，让建筑呈现出变幻多彩的形态效果。

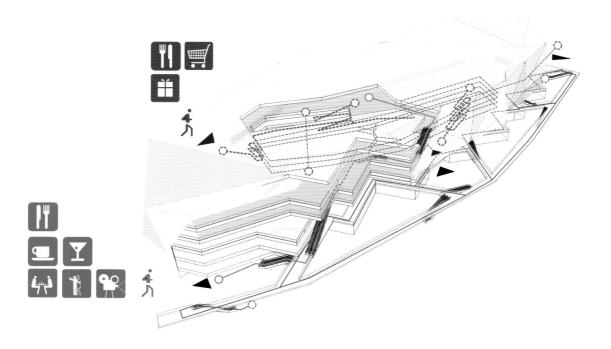

Combined Circulation
联合循环系统

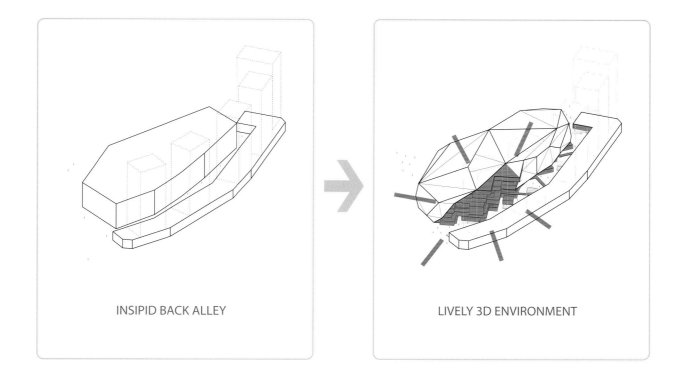

INSIPID BACK ALLEY

LIVELY 3D ENVIRONMENT

Conceptual Diagram
概念图

Two complementary pedestrian routes energize the dynamic shopping experience at Thaihot Mall Fuzhou. A 12-hour day route and a 24-hour route, retail programs that operate during the day are consolidated into a 12-hour shopping podium, circulation routes and terraces are carved out of this podium facilitating access to the rooftop. The rooftop offers a variety of activities such as miniature golf and fine dining, it is accessible by the 24-hour route. This route is further animated by its adjacency to the pedestrian street.

The north façade becomes a living organism full of movement and energy, Thaihot Mall Fuzhou is set to become a unique destination full of surprise and discovery and distinctive reflection on the future of city lifestyle retail.

为了方便顾客自主购物，享受便利的购物体验，广场内设有两条互补线路，一条营业时长达 12 小时，另一条营业时长达 24 小时。白天营业的零售业务形成了 12 小时开放的购物平台，供人们选择任意线路进行购物。此外，两天线路和步行楼梯都可直达楼顶。顶楼设置了丰富的娱乐设施，例如迷你高尔夫游戏和精致的餐厅。楼顶设施是 24 小时营业的，所以你可以选择 24 小时皆营页的线路自由上下。其实，相比 12 小时营业线路，24 小时营业线路灵活力更强，活力更佳。

本案被认定为是独一无二又令人惊喜的创新方案，其北面是充满动感和能量的健身活动区，整个建筑体现了零售业务独特的经营模式，反映了城市未来零售业的发展目标。

IMPORTANNE CENTER SARAJEVO

萨拉热窝 Importanne 购物中心

Architect: Studio Non Stop
Client: Teloptic doo Sarajevo
Location: Sarajevo, Marijin dvor
Site Area: 10,000 m²

设计公司：Studio Non Stop
客户：Teloptic doo Sarajevo
地点：萨拉热窝 Marijin Dvor 区
占地面积：10 000 m²

STRUCTURE AND MATERIAL 结构与材料

STRUCTURE
Reinforced Concrete Structure

MATERIAL
Façade Panels, Glass

结构
钢筋混凝土结构

材料
立面板、玻璃

Importanne Center is formed of eight towers which are connected in between with vertical gardens and so called façade bites which represent a symbolic place of connection of individual buildings forming the structure of urban puzzles.

The buildings are defined as a multifunctional business center that includes office space, hotel, residential, shopping center and underground garage. According to its urban and programmatic context the building is formed of eight stick elements - the towers, which are compressed into a single unit forming multi-block. Each tower is still functional and a formal separate entity which is connected with the other towers in a way to form a vertical garden façades.

Importanne 中心共由八座塔楼组成，各塔楼之间以垂直花园相连，被戏称为"咬伤的面孔"，象征着构成整个城市版图的个体建筑间的联系。

建筑物被定义为多功能的商业中心，内含办公区、酒店、住宅、购物中心和地下停车场。按照城市规划和建筑方案，整个建筑由八个重要塔楼构成，每个塔楼被简化成一个单元。各塔楼具有不同的功能，彼此之间通过一个垂直花园连接着。

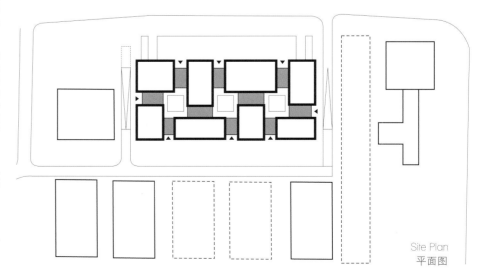

Site Plan
平面图

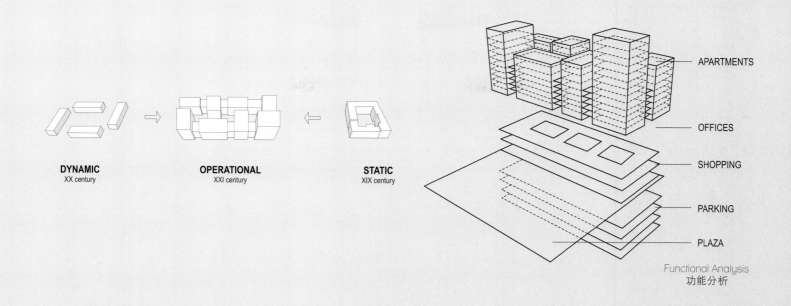

Functional Analysis
功能分析

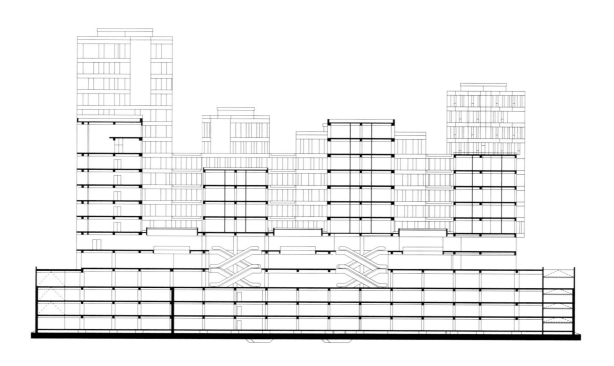

Section A-A
剖面图 A-A

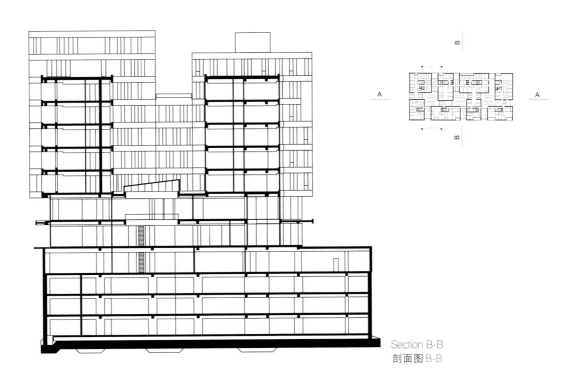

Section B-B
剖面图 B-B

Thanks to the vertical and horizontal circulation system, the facility is provided with a range of functional interconnection in accordance with the market conditions and tenant requirements.

Each of the towers has its own special aesthetic expression through the rhythm of the façade openings, colour of the façade panels, as well as the position of the so-called façade bites. In this way, a very interesting and dynamic interconnection, interaction and communication between the separated towers of urban multi-block is achieved.

得益于建筑物内的垂直和水平的循环系统，该建筑可按照市场条件和租户要求实现设施的各种功能互动。

通过立面开口的错落、立面板的色彩、以及所谓立面连接的方式，不同的塔楼展现了不同的美感，同时实现了多功能区域内各个独立塔楼之间的互动和沟通。

Greenery Compressing
浓缩绿化

SHAPE ANALYSIS 造型分析

ECOLOGICAL SHAPE
As a symbolic building, Importanne Center constituted by eight towers and connected with a vertical garden. And through façade openings, vertical panels color shows a different beauty.

生态造型
作为一个具有象征性意义的建筑，Importanne 购物中心由八座塔楼构成，并以垂直花园相连，主要通过立面开口、立面板的色彩展现出不同的美感

6th Floor Plan
六层平面图

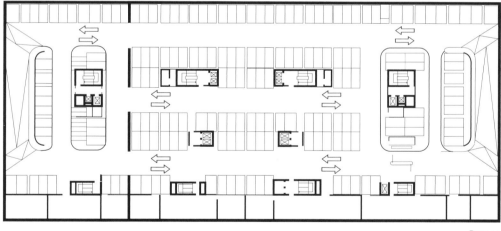

2nd Floor Plan
二层平面图

Garage
停车场

SKIN ANALYSIS 表皮分析

LIGHT SKIN

The building shows different beauty through the vertical panel colors outside to realize the interaction between the separated towers of urban multi-block.

轻表皮

建筑通过其外部的色彩来展现不同角度的美感，实现多功能区域内各个独立塔之间的互动。

COMMERCE AND OTHERS
商业展示及其它

ROCA LONDON GALLERY 乐家伦敦展廊

Architect: Zaha Hadid Architects
Client: Roca Ltd
Location: London, UK
Site Area: 1,100 m²

设计公司：扎哈·哈迪德建筑事务所
客户：乐家卫浴公司
地点：英国伦敦
占地面积：1 100 m²

STRUCTURE AND MATERIAL 结构与材料

STRUCTURE
Steel Framed Structure

MATERIAL
Glass Reinforced Concrete, Glass Reinforced Gypsum, Resin and Tiles

结构
钢架结构

材料
玻璃钢筋混凝土、玻璃纤维石膏、树脂、瓷砖

Zaha Hadid Architects, acclaimed winners of this year's Stirling Prize, have designed the new Roca

London Gallery located in the well-known Chelsea Harbour district near King's Road.

The Roca London Gallery expresses Roca's commitment to innovative design, sustainability and wellbeing through an architecture that offers visitors a unique experience which is both visually and interactively engaging. Taking its inspirational cue from the idea of a space created by water in its various different states, ZHA's bespoke design defines a flowing and porous space for the 1,100 m² Roca London Gallery.

The design brings about a connective language between the architecture and the bathroom products, with the movement of water 'carving out' the interior and moving through the Gallery as individual drops. A flowing, all-white space made of faceted GRG (gypsum) panels serves as a central axis of the Gallery. Around this a number of smaller connected semi-enclosed spaces can be viewed through openings in walls. As a result, the visitor never feels enclosed in one space, but can always see beyond it into the space through overlapping and cutaway forms that enable a pleasing permeability to the Gallery.

负责本案的设计公司是荣获今年的斯特灵建筑设计大奖的扎哈·哈迪德建筑事务所，它是时下最炙手可热建筑事务所。

最近，这家公司刚刚完成了坐落于伦敦切尔西海港区英皇道的新乐家展廊设计项目。

通过本案，设计师展现了乐家的创新、环保和发展，为人们打造出一种独一无二的视觉享受和亲密的互动体验。在千变万化的水流形态中，设计师获得了本案的设计灵感，在占地1 100 m²的规划用地上打造了一个多孔分割、液态流动的展廊。

既然本案的设计要以"流动之水"为主题，就不可避免地要在设计构思中将建筑与卫浴产品联系了起来：汹涌的水流冲刷而入、沿着室内展廊流淌、最终汇聚成一个个小湖。流动着的纯白空间就是展廊的主轴，这是采用石膏嵌板镶嵌的。围绕这个主轴，透过可视见的墙壁开口，就能看到多个相互连通的小型半封闭空间。游客并不感觉自身被困在一个封闭的空间里，因为他们会发现空间与空间之间是层叠的。值得高兴的是，游客果然更喜欢这种不封闭的自由空间。

Water defines the landscape of the interior space, creating a sense of mobile liquidity reinforced by a series of elongated, illuminated water drops. These cascade around the ceiling as a set of lighting fixtures, down the walls as shelves for books, media and small products, and onto the floor as tables and seating. Their fluid lines of convergence both lend each area of the space an individual identity and connect them by the way they define a feeling of movement.

为了很好地诠释 "流动之水" 这一主题，室内尽是一连串细长的水流或者一滴滴发光的水珠。但是天花板只安装了一组灯具，并且在依墙之处附有可放书本和小型展览品的架子，地面也安放了一些椅子。虽然装饰不多，但是它们的线条十分流畅，呈现出动态美感，非常符合主题。此外，这种辅助装饰还能更好地将各个空间联系起来。

SHAPE ANALYSIS 造型分析

DIGITAL SHAPE

The gallery takes the flow of water as prototype. As a result, the entire interior space is portrayed as the ever-changing water, mutual penetrated and interlinked.

数字造型

本案将水的流动形态作为设计原型，把整个室内空间塑造成多变的流水形态，每种流水形态都相互交织，自相融汇。

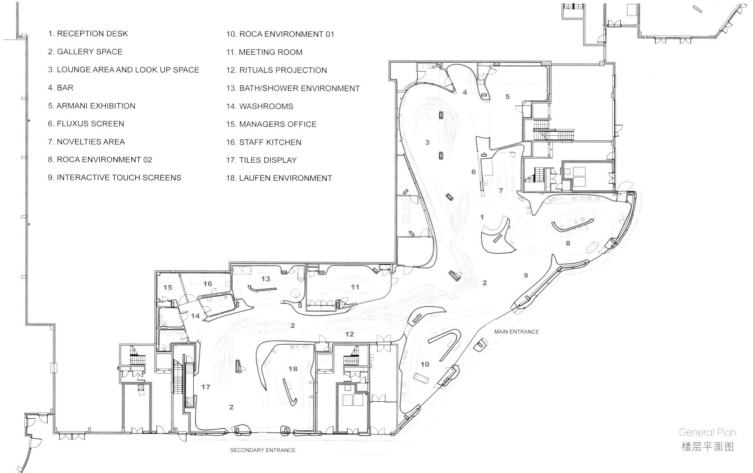

1. RECEPTION DESK
2. GALLERY SPACE
3. LOUNGE AREA AND LOOK UP SPACE
4. BAR
5. ARMANI EXHIBITION
6. FLUXUS SCREEN
7. NOVELTIES AREA
8. ROCA ENVIRONMENT 02
9. INTERACTIVE TOUCH SCREENS
10. ROCA ENVIRONMENT 01
11. MEETING ROOM
12. RITUALS PROJECTION
13. BATH/SHOWER ENVIRONMENT
14. WASHROOMS
15. MANAGERS OFFICE
16. STAFF KITCHEN
17. TILES DISPLAY
18. LAUFEN ENVIRONMENT

MAIN ENTRANCE

SECONDARY ENTRANCE

General Plan
楼层平面图

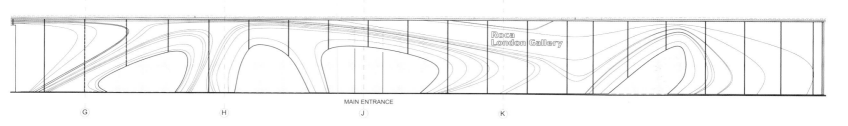

East Elevation
东立面图

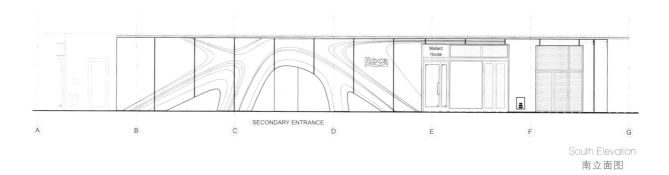

South Elevation
南立面图

All the panels, which are made of GRC, or fibre reinforced concrete and extend up to 2.20 metres in height, have been pre-fabricated in moulds and constructed on-site. The façade is made of 2 x 4 metre panels of 800kg each. The furniture is made from GRP, or reinforced plastic, including the cove-shaped reception desk. The lighting scheme created by Isometrix is also innovative in a complementary way, with special features including washing the walls in light and a mix of direct and dispersed mood lights. A special feature of the Gallery is the floor of the product exhibition areas, which has a mosaic of porcelain tiles designed exclusively for the space by Zaha Hadid Architects. With each one cut and laid individually, the design creates an optical effect inspired by a water current.

所有的面板都是在建筑现场使用模具特制而成的。本案采用玻璃纤维增强混凝土，增强面板的坚硬度，还使用了纤维制品材质，以保证面板长度达到2.2 m。外立面面板的规格统一为2 m×4 m，荷载0.8 t。展廊内的家具，包括半月形接待台等，都使用了坚硬度极高的玻璃钢或增强塑料材料。Isometrix公司的照明方案为本案增色不少。它的创新之处在于采用互补式照明方案，不但使用壁灯打亮内墙，还使用直射灯与散射灯相互搭配，营造一个忽明忽暗、带有神秘感的展廊。此外，本案最亮眼的设计当属陈列区的地板图案。地板使用本案设计事务特别打造的马赛克瓷砖。这些瓷砖的花纹是独一无二的，尺寸也各不相同。大大小小的瓷砖拼凑出一副诗意般的"流水"画面，不但呼应了主题，烘托出整体环境，还体现了设计师精巧的构思。

SKIN ANALYSIS 表皮分析

DIGITAL SKIN

The design theme of water movement extends to the dynamic façade of the Roca London Gallery, which appears initially to the visitor approaching the architecture like a set of ripples in movement across the exterior of the ground level space. The grey façade has large apertures for the main entrance and windows and an appearance of tactility, creating a sense of intrigue on the street as the visitor approaches.

数字表皮

本案将"流动之水"这个主题延伸到建筑层面上，设计出带有动感的立面。其外部空间形似一组浮动的涟漪，让游客刚走进建筑物，就被浮动的波纹造型所吸引。除此以外，展廊的主入口、窗口以及触感都极力模仿水流的形态。它们被粉刷上灰色油漆，形状阔大，触感十分柔滑。这种出色的流水立面，吸引了人们的注意力。

THE CULTURE YARD

文化庭院

Architect: AART Architects
Location: Elsinore, Denmark
Site Area: 13,000 m²
Photography: Adam Mørk

设计公司：AART 建筑事务所
地点：丹麦艾尔西诺
占地面积：13 000 m²
摄影：Adam Mørk

STRUCTURE AND MATERIAL 结构与材料

STRUCTURE
Concrete Frame

MATERIAL
Steel, Glass, Aluminium

结构
混凝土框架结构

材料
钢材、玻璃、铝材

In many years the attention has been aimed at the site adjacent, where the UNESCO World Heritage site, Kronborg Castle, which is famous for its role in Shakespeare's *Hamlet*, exerts its magnetic pull on both tourists and local citizens of Elsinore – but now Elsinore's old shipbuilding yard has been transformed into a 13,000 m² cultural and knowledge centre, including concert halls, show rooms, conference rooms, a dockyard museum and a public library. The Culture Yard symbolizes Elsinore's transformation from an old industrial town to a modern cultural hub. In this way, the yard is designed as a hinge between the past and present, reinforcing the identity of the local community, but at the same time expressing an international attitude, reinforcing the relation between the local and global community.

The contrast between past and present permeates the Culture Yard. For instance, the original concrete skeleton with armoured steel has been reinforced, but left exposed as a reference to the area's industrial past. The historic context has thus been the main structural idea in the design process, ensuring the keen observer will discover a chapter of history in every corner of the yard and every peeling of the wall. In other words, if you want to understand what Elsinore really is, what the intangible blur between past and present feels like, this is the place to visit.

多年来本案的邻近场地都备受关注，因为那里是联合国教科文组织认证的世界遗产——卡隆堡宫城堡的所在地。卡隆堡宫城堡因为莎士比亚的著名戏剧《哈姆雷特》而闻名天下，它对来自世界各地的游客和艾尔西诺本地居民都有着不可言喻的吸引力。但现在埃尔西诺的老造船厂已被改造成占地 13 000 m² 的文化知识中心，里面设有音乐厅、展示厅、会议室、船厂博物馆和公共图书馆。可以说，文化庭院的改建象征着埃尔西诺从老工业城市到现代文化中心的转变。因此，文化庭院被设计成连接过去与现在的交互场所。它既要强调其在当地文化中不可触动的重要身份，同时又要表达其对国际历史文化的态度，再者，作为桥梁促进本土文化与国际社会之间的关系。

过去和现在两种对比式设计渗透到文化庭院的各个角落。例如，将早已被加固好的钢筋混凝土框架又裸露在外，引导游客想起该地区的工业时代。本案将老造船厂所承载的历史背景作为构思主线贯穿整个设计进程，确保敏锐的游客能从庭院的每个角落里或从每堵颓垣残壁中发现工业时代的痕迹。换句话说，如果游客想了解真正的埃尔西诺，或者想感受过去和现在的模糊界限，那么文化庭院就是最好的去处。

Thanks to architectural features such as wrought iron stairs and concrete elements, interacting with modern glass structures and interior designs, the contrast between the days of yore and the present becomes evident. It is the Culture Yard's way of playing with the field of tension between old and new, making the notion of past versus present, the industrial society versus the information society, constantly present.

因于工业时代那些锻铁楼梯和混凝土构件与现代社会那些玻璃材料和室内设计元素各自突出,使得往昔与现在的对比更为明显。文化庭院以它独有的方式协调新旧之间的冲突,让过去与现在、工业社会与信息时代并存。

SHAPE ANALYSIS 造型分析

DIGITAL SHAPE

The Cave and the Pyramid are examples of how the design, function and construction form a whole. The Cave is designed as a hanging auditorium that protrudes 15 meters out of the building above the main entrance, thus providing a dramatic effect and. The Pyramid protrudes 9 meters out of the eastern building and is an extension of the architectonic lines of the façade. The extension of lines forms a horizontal pyramid shape, resulting in the name the Pyramid.

数字造型

本案的洞穴设计方案和金字塔造型都是出色的模范案例，教你如何设计、运作直到最后建设成一个整体。洞穴被设计成悬挂式礼堂，悬浮在本案主入口上方 15 m 处，以其所处位置独特而颇引人注目。同样，位于本案东部的金字塔建筑也向外突出 9 m，如线一样延长了本案的外立面。由于延长线形成了一个水平锥形，因此得名为"金字塔"。

SKIN ANALYSIS 表皮分析

TRANSPARENT SKIN

The transparent façade that encloses the old shipbuilding yard gives a glimpse of the historic buildings. The façade gives the Culture Yard an intriguing identity where the distinctive steel elements of the construction are used as a reference to the area's original function as a shipyard. Furthermore, the façade has an environmentally friendly function, as the south facing parts of the façade is covered with perforated aluminium plates that provide shade during the day.

透明表皮

透明立面将老造船厂围合起来，人们可以通过透明立面惊鸿一瞥这座沧桑的历史建筑。而且透明立面赋予本案有趣的身份象征，那些独特的钢构件结构足以让游客明晓它的前身就是一家造船厂。此外，立面还具有环保功能，朝向南方的立面被装上穿孔铝板，在白天里起着遮阳的作用。

AUTOMOTIVE CENTRE OF EXCELLENCE

卓越汽车中心

Architect: Lyons
Location: Victoria, Australia
Site Area: 5,000 m²

设计公司：Lyons
地点：澳大利亚维多利亚省
占地面积：5 000 m²

STRUCTURE AND MATERIAL 结构与材料

STRUCTURE
Reinforced Concrete Structure

MATERIAL
Steel, Concrete, Glass, Brick

结构
钢筋混凝土结构

材料
钢材、混凝土、玻璃、砖材

The new Automotive Centre of Excellence (ACE) in Melbourne's Docklands accommodates a dedicated training and showcase facility for Australia's automotive trades and manufacturing. It consists of high-bay workshop spaces, specialist workrooms, classrooms and office accommodation.

A strategy was needed to develop an appropriate civic scale for this small public building in the context of its surrounding commercial urbanscape.

新建成的卓越汽车中心（ACE）将成为墨尔本一个专为澳大利亚的汽车贸易和制造业服务的专业培训基地和汽车展示场所。本案设有高顶棚工作坊、专家工作室、学员教室和办公室。

在宏观的城市环境中，本案设计师需要为这座小型公共建筑打造适合的形象，使它与周围的商业景观相结合。

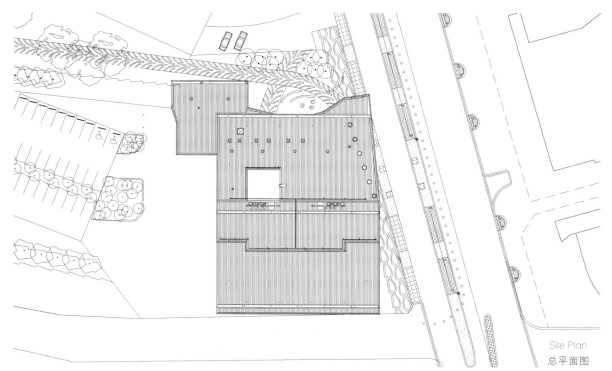

Site Plan
总平面图

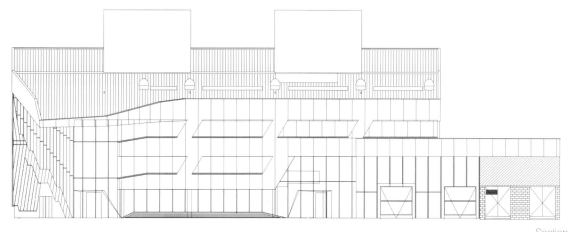

Section
剖面图

The building also absorbs sources from automotive culture, and its relationships with the city; kerb signs, tyre treads, city overpasses, and the sheen of car showrooms. The interiors evoke something of the automotive predilection for contrasting the technological and mechanical with the finished and the smooth.

The main foyer with its monumental staircase acts as the key circulation pathway through the building. Visitors experience a transition from traditional technical college materiality; raw blockwork, exposed steel and concrete to contemporary applications of carbon fibre and glass projection technology.

此外，本案设计还融汇了汽车文化，添加些许墨尔本特色——街边的石刻指示牌、轮胎式路面、城市立交桥和汽车展厅的灯光效果等。其中，室内设计就能引起汽车爱好者的共鸣，精湛的汽车技术让人惊诧连连，平滑的机械触感。

带有大型楼梯的大堂是本案的主通道。游客从这里出发，能观看传统工学材料转型到现代工业材料的发展历程，从原料砖造物、裸露的钢筋和混凝土一直演变到碳纤维、玻璃投影技术等现代工学材料。

SHAPE ANALYSIS 造型分析

OPEN SHAPE

We looked at the history of the Docklands to identify a gesture to allow the building to compete with its high rise neighbours – in particular the history of the "big shed", evident in the adjacent railway sheds. The roof is a large, simple gable which connects ACE to other Melbourne based industrial training spaces.

开放造型

本案设计师回顾了墨尔本的发展历程，希望找出一种特殊的元素，建设一栋有别于周边高层大楼的建筑，尤其要比相邻铁路的"大型棚盖"更加亮眼。本案屋顶是一块巨大而简洁的三角棚盖，其造型能将本案与墨尔本其他工业训练场联系在一起。

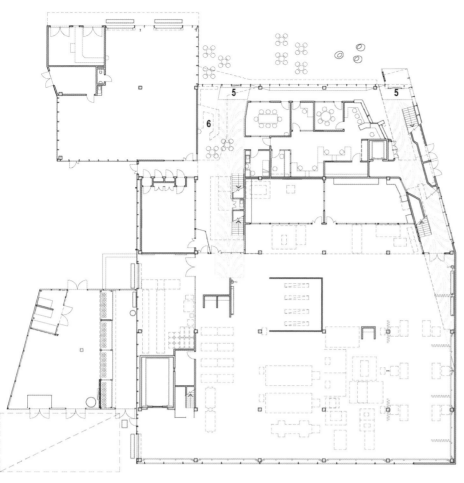

Ground Floor Plan
首层平面图

SKIN ANALYSIS 表皮分析

ENERGY SKIN

The shed façade system incorporates automated louvres which enable the workshop spaces to be naturally ventilated. The offices and classroom spaces are cooled by an active thermal mass system. In combination with other environmental sustainable design features the building has achieved a 5-Star Green Star environmental rating.

节能表皮

网格立面系统所采用的自动百叶窗能帮助室内车间自然通风,办公室和学员教室所采用的热容量系统能调节室温,此外还有其他有利于环境可持续发展的设计,这些元素促使本案成功荣获五星级绿色环境评级。

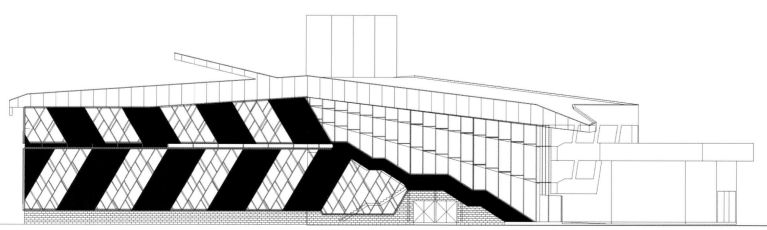

Section
剖面图

维特拉展馆

Architect: Herzog & de Meuron
Client: Vitra Verwaltungs GmbH
Location: Germany
Site Area: 12,349 m²
Photography: Leon Chew, Iwan Baan, Vitra
Drawings: Herzog & de Meuron

设计公司：Herzog & de Meuron
客户：Vitra Verwaltungs GmbH
地点：德国
占地面积：12 349 m²
摄影：Leon Chew、Iwan Baan、Vitra
图纸：Herzog & de Meuron

STRUCTURE AND MATERIAL 结构与材料

STRUCTURE
"Stacked" Structure

MATERIAL
Glass, Concrete

结构
"堆叠"结构

材料
玻璃、混凝土

本案借鉴了工业化生产中的"堆叠"、"挤延"和"压缩"手段，将简单的整体配置成多个复杂且内外相互连通的空间，形成了一种仿若多座"长方形房子"堆叠而成的奇妙的建筑造型。

SHAPE ANALYSIS 造型分析

PICTOGRAPHIC SHAPE

By stacking, extruding and pressing – mechanical procedures used in industrial production – simply shaped houses become complex configurations in space, where outside and inside merge. Eventually stacked blocks of rectangular houses form the project

象形造型

本案借鉴了工业化生产中的"堆叠"、"挤延"和"压缩"手段，将简单的整体配置成多个复杂且内外相互连通的空间，形成了一种仿若多座"长方形房子"堆叠而成的奇妙的建筑造型。

Over the past few years, Vitra has acquired a wide-ranging Home Collection. The quantity and variety of objects by many different designers led to the idea of building a showroom to present the items to the public. There would also be additional space to be used as an exhibition venue for selected parts of the collection or even as an extension of the Vitra Museum itself. A shop, a cafe linked to the outside and conference rooms complete the program.

The "Vitra Haus" is a direct, architectural rendition of the ur-type of house, as found in the immediate vicinity of Vitra and, indeed, all over the world. The products that will be on display are designed primarily for the private home and, as such, should not be presented in the neutral atmosphere of the conventional hall or museum but rather in an environment suited to their character and use.

在过去几年里，Vitra 家居公司珍藏了许多出自名设计师之手的家具收藏品。这些收藏品数量繁多、种类齐全，让 Vitra 公司萌生了建造展馆来陈列这些收藏品的想法。除了公共展览厅，本案还设置了指定展品的展览空间和类似 Vitra 博物馆的展厅。此外，为了丰富并完善项目规划，展馆外部还配套有商店、咖啡厅和会议室。

Vitra Haus 的设计概念包含了两个主题：原型房子和堆积体量。事实上，本案与邻近 Vitra 的周边房屋造型、甚至是世界各地的房屋造型都有相似之处。在此展出的家居产品是专为私人住宅设计的，它不适宜摆放在传统大厅或博物馆中展示，只适宜放在私宅中。私人住宅的主人可根据它们的特点和功能选择适合的家具。

Location
区位图

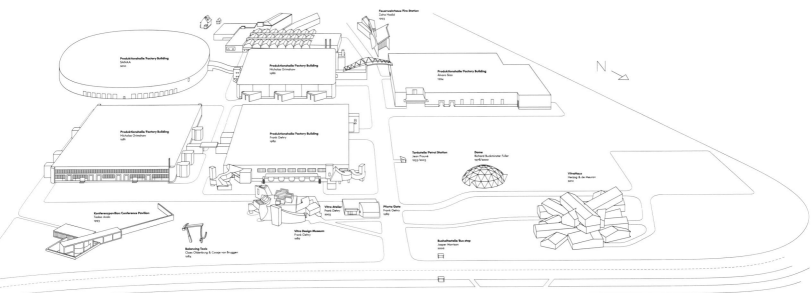

Aerial Plan
规划鸟瞰图

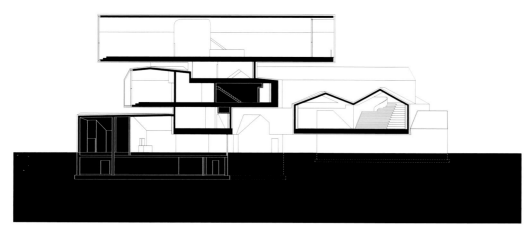

Section S1
剖面图 S1

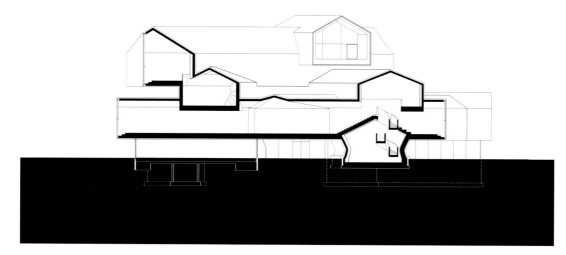

Section S2
剖面图 S2

SKIN ANALYSIS 表皮分析

LIGHT SKIN

The thick black coating expresses a deep sense, while internal warm environment is revealed from the front transparent glass façade. All that gives a overall effect that the building in the dark, but revealing bright to outside world.

轻表皮

质厚而色黑的立面涂层所传递出深冷沉稳的感觉，与前面的玻璃外墙形成强烈对比。透明的玻璃将和暖的室内光线流露于表，在夜幕下观望，打造出一种漆黑里泛出暖光的视觉效果。

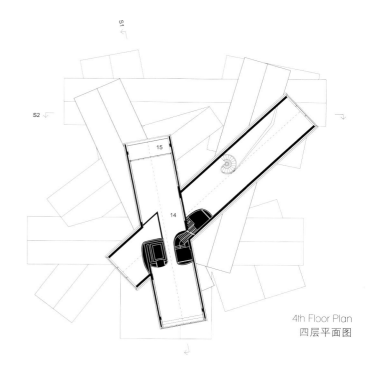

4th Floor Plan
四层平面图

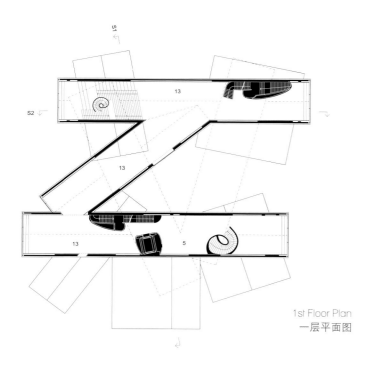

1st Floor Plan
一层平面图

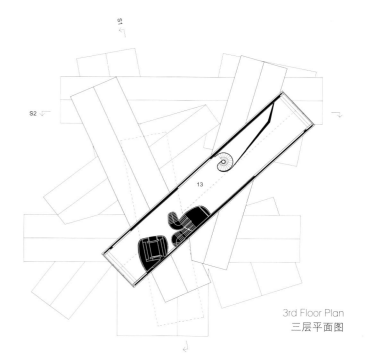

3rd Floor Plan
三层平面图

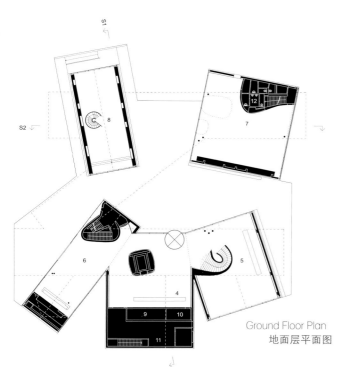

Ground Floor Plan
地面层平面图

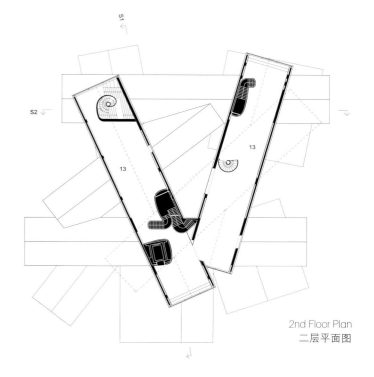

2nd Floor Plan
二层平面图

SHERBROOKE EXHIBITION CENTER 舍布鲁克展览中心

Conception Architect: CCM² - Côté Chabot Morel Architects
Realization Architect: ArchiTech Design inc.
Client: Ville de Sherbrooke
Location: Sherbrooke, Québec, Canada
Built Area: 9,570 m²
Photography: Stéphane Groleau / stephanegroleau.com

设计公司：CCM² - Côté Chabot Morel 建筑事务所
施工单位：ArchiTech Design inc.
客户：舍布鲁克政府
地点：加拿大魁北克舍布鲁克
建筑面积：9 570 m²
摄影：Stéphane Groleau / stephanegroleau.com

STRUCTURE AND MATERIAL 结构与材料

STRUCTURE
Steel Structure

MATERIAL
Aluminium Panel, Steel Cladding

结构
钢结构

材料
铝面板、钢覆层

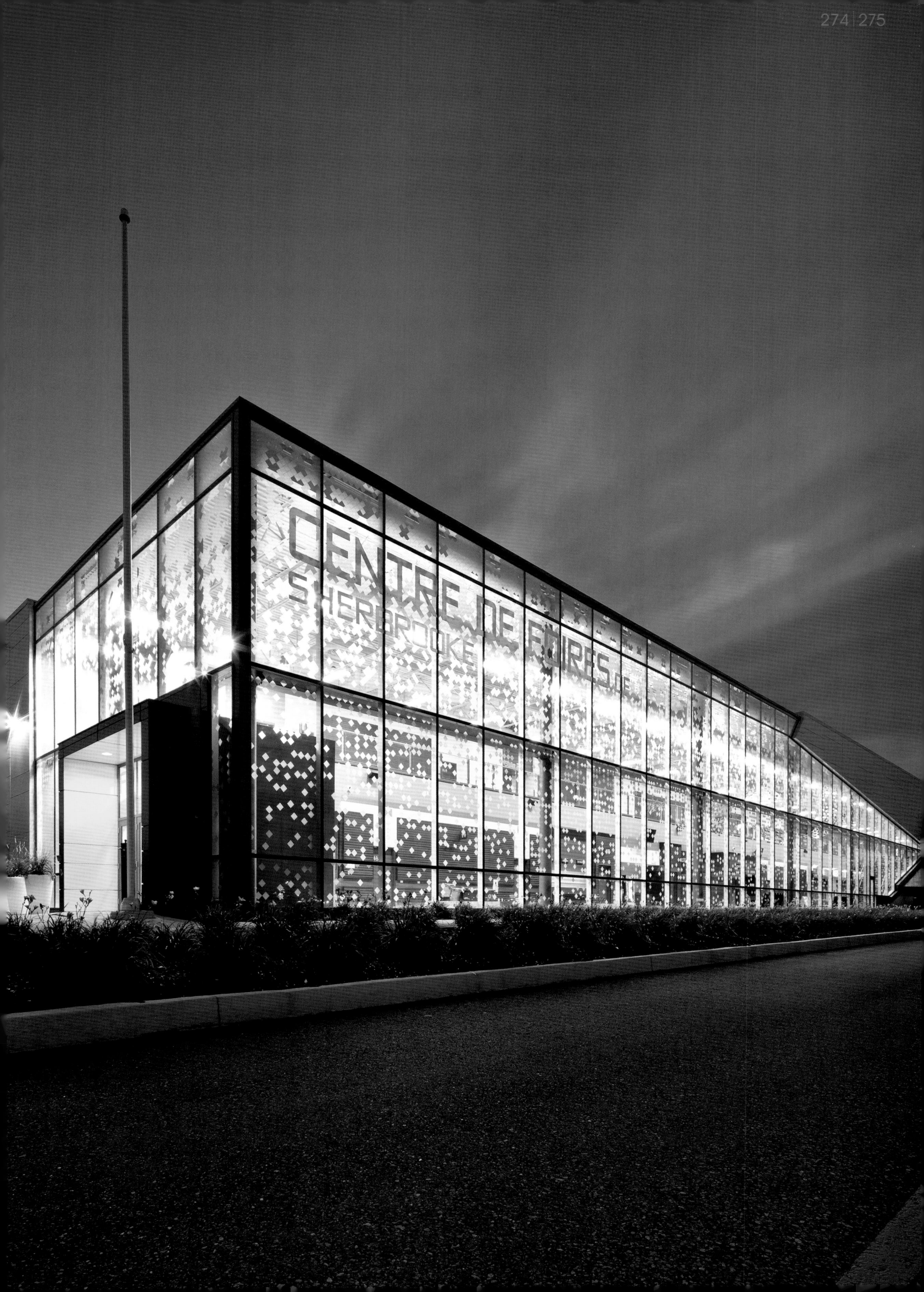
CENTRE DE FOIRES DE
SHERBROOKE

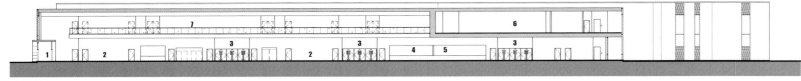

1- Entrance
2- Hall
3- Entrance to exhibition space
4- Coatroom
5- Ticket office
6- Open space
7- Futur meeting rooms

Section
剖面图

SHAPE ANALYSIS 造型分析

OPEN SHAPE
The lower part of the building is enclosed by a glass façade with thick lines depicting a double-floor entrance. The composition with an overall regular shape makes the Exhibition Center a unique venue for the visitors of the exhibition centre.

开放式造型
本案低层部分以玻璃立面作围合,粗犷的线条勾勒出双层玻璃入口。这座方正的建筑物,让进场参观的游客无一不感觉到本案更像是特色场馆,而非展览中心。

Located near the downtown of Sherbrooke, the Exhibition Center serves as a gathering place and an unique representation place in the region, Which offers a widerange of opportunities in terms of rental space.

The project's main objective is to create divisible exhibition areas if needed for dedicated events for the wide public. The program includes the creation of exhibition halls, meeting rooms, offices, central hall, cloakroom, ticket office, restaurant and other technical spaces.

本案位于舍布鲁克的闹市附近,以独特的造型被选为舍布鲁克市标志性建筑,也因为可租赁各种不同类型的场地而成为市区重要的集贸市场。

本案设计的目标是打造出能满足社会各种展览需求的、可分割的灵活型展览区。项目的规划内容包括建造大型特色展厅、会议室、办公室、中心大会堂、盥洗室、售票处、餐厅和其他技术控制室。

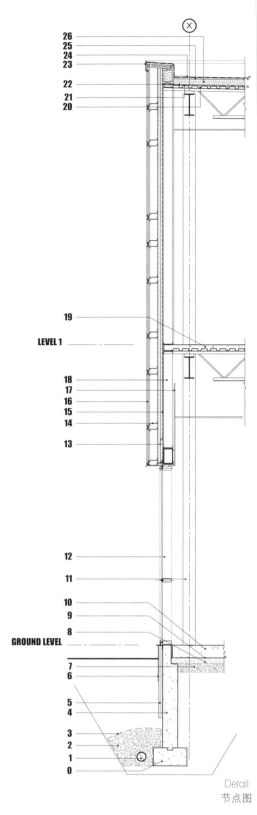

Detail
节点图

SKIN ANALYSIS 表皮分析

LIGHT SKIN

The upper area covered by white aluminium panel and the lower glass façade create a subtle visual effect. The glass façade reveals interior activities to the outside to presented clearly the theme of exchange and interaction in the Exhibition Centre.

轻表皮

本案高层部分所覆盖的白色铝板和低层用以围合的玻璃立面相互映衬，打造出若隐若现的视觉效果。这种用玻璃立面让室内的活动情况完全暴露在外的设计，清晰明了地表达展览中心的设计主题——"交流与互动"。

The total project area is approximately 103,000 square feet. The main exhibition hall of 60,000 square feet, designed according to the highest standards in terms of functionality and specialized equipments, can be divided into three exhibition spaces when various events take place simultaneously. On each sides of the exhibition spaces are located the related services. On one side, the technical spaces with the loading areas, storage areas and recycling areas, and on the other side, public spaces including the lobby, cloakroom, ticket office, services, restaurant and meeting rooms on the second floor.

The architectural concept can be resumed essentially by the desire to express the internal activity of the building. Upon his arrival on the site, the visitor sees a large glass hall that show partially the internal activity of the exhibition center. The main hall contains the essential services and allows visitors to walk through a space flooded by natural light. The pixels pattern on the glass surface created by a semi-transparent film adds a light veil on the indoor activities taking places in the hall and adds a subtle texture to the glass wall. From the outside, the angular shape rises and highlights the main entrance. Sculpted and defined as a break of the glass block, the main entrance is the key element of the project and marks the transition between the exterior and the interior environment. The use of bright yellow color shows the internal activity of the exhibition center. This color sequences animates the façades like if the internal boiling energy was breaking up partly and shows the festive character of the place.

本案总面积约 9 570 m²，展览区占地 5 574 m²，其功能和设备都按照最高标准来设计。展览区灵活分割成三个独立的展馆，以满足不同专题同时展览的需要。各个功能区域都有相应的服务设施。建筑物的一边是负责装卸、储存和回收技术的控制室，另一边则是包括大堂、盥洗室、售票处、咨询服务部、餐厅以及二楼会议室在内的公共活动空间。

从本质上来说，本案的设计理念就是展示开放式室内空间。因此，当游客抵达现场就会被眼前一座巨大透明的玻璃大厅所吸引。整个展览大厅无论室内还是室外都被一览无余。主展区提供完善的服务设施和优美的环境，当太阳之光散落满地之时，游客便可踩着细碎的光影漫步展区。透过玻璃幕墙看去，展厅内人们活动的身影就轻盈地映射在玻璃墙上，在玻璃表面构成了妙曼的图案，难道玻璃幕墙不像放映着一部半透明的影片吗？综观展厅外部，主入口的棱角造型张扬而出色。方正有角而精雕细琢的造型主入口犹如玻璃幕墙的一块缺口，是本案的关键，起着转换室内与室外的作用。其中标志由室外转换到室内的立面，采用了亮黄色来增添内墙生气，让人感到热血沸腾，同时也增添本案的节日气氛。

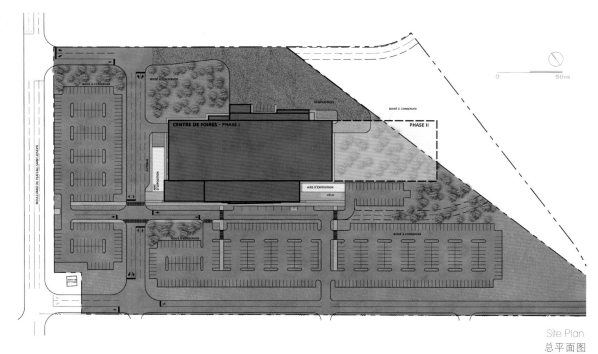

Site Plan
总平面图

1 - Entrance
2 - Hall
3 - Entrance to exhibition space
4 - Coatroom
5 - Ticket office
6 - Exhibition space
7 - Technical area

Ground Floor Plan
首层平面图

热那亚配料展览会 B 展馆

Architect: Ateliers Jean Nouvel
Client: Fiera di Genova
Location: Genova, Italy
Built Area: 36,000 m²
Photography: Philippe Ruault

设计公司：让·努维尔建筑事务所
客户：Fiera di Genova
地点：意大利热那亚市
建筑面积：36 000 m²
摄影：Philippe Ruault

STRUCTURE AND MATERIAL 结构与材料

STRUCTURE
Reinforced Concrete Structure
MATERIAL
Glass, Concrete, Steel

结构
钢筋混凝土结构
材料
玻璃、混凝土、钢筋

Site Plan
PLANIMETRIA GENERALE 总平面图

SHAPE ANALYSIS 造型分析

OPEN SHAPE

The ground floor separates from outside by glass informally, in fact, it opens towards the port completely. The volumes also extend to the harbor, trying to blend into it.

开放式造型

建筑的底层只是以玻璃作形式上的分隔，实为向着港口完全开放。建筑的体量也朝着海港的方向延伸，试图与海港融为一体。

It's simple. Too simple to be so simple. It's first and foremost a plane, a blue plane, too blue, that sky blue of old underexposed photos or of early cinemascope. The plane is a roof. It can be seen from the city, from the road beside it, as an abstraction, an immense blue mirror in which the blue of the sky, or the gray of clouds which turn blue beneath the rain, is reflected.

On the inside it's not much more complicated, at least not on first sight. Two exhibit halls open up to the sea. One, under the roof plane, is a covered terrace opening onto the port. The other is simply the extension of the quayside into the building; it also is open to the port. When I say open I mean entirely glazed, but also operable so that it be opened when wind and temperature conditions permit. Both halls are placed under great mirrors that evoke the surface of the sea. One is fashioned of bumps or wavelets like the waters of a port that lap at a dock. The other is fashioned of long sinusoidal waves parallel to the quay whose amplitude decreases as they approach the port. The mirrors diffract and re-form the image of the exhibits they cover and, to a lesser degree, the image of the boats in the port...

再没有比本案更为简单的设计了。首先是它的平面、碧蓝如海的平面，就像是曝光不足的老照片或早期银幕中的蔚蓝天空。该平面就是建筑的屋顶，从城市和旁边的街道上看去，这屋顶就像是一面抽象的巨大蓝镜，反射出蓝色的天空或下雨前的乌云。

建筑的内部结构也不是非常复杂，至少乍看之下不会有复杂感。两间展览大厅皆采用开放式设计，面朝大海，位于平面屋顶的下方的一间为通向港口的露台提供阴凉；另一间是码头到建筑内部的延伸，也同样面朝港口。在这里，所有的玻璃都是开放设计的、在风力和温度条件允许的情况下，可将玻璃窗户打开。两间大厅都位于镜面屋顶下方，让然联想到平静的海面。其中一间大厅被塑造成凸起或微波状，就像是扑打到码头上的水浪；另一间则打造成长长的正弦波、与码头平行，其幅度朝着码头方向逐渐缩小。镜面般的屋顶衍射并重新形成其下方的大厅形象，就像是一艘在港口的船。

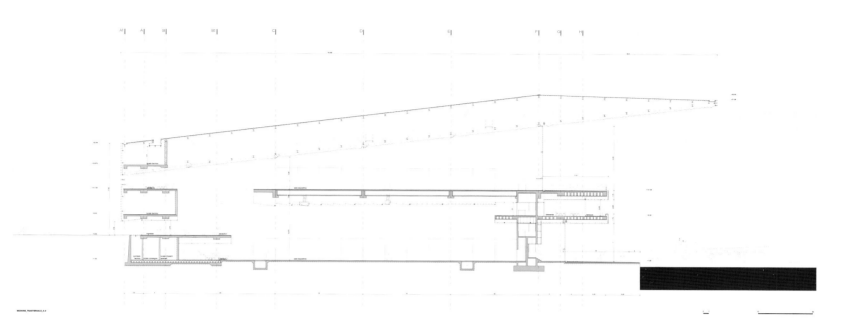

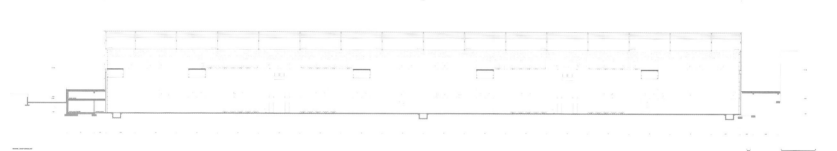

Side Section
侧剖面图

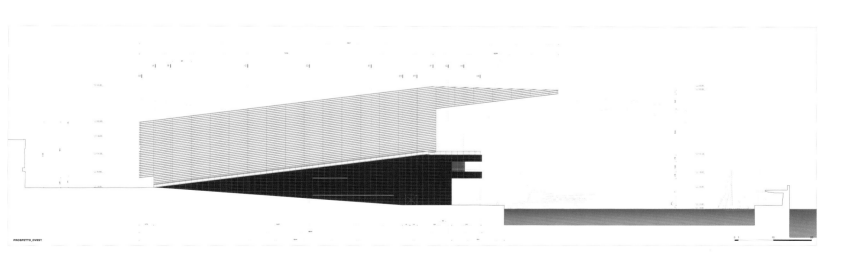

Structure Analysis
结构分析图

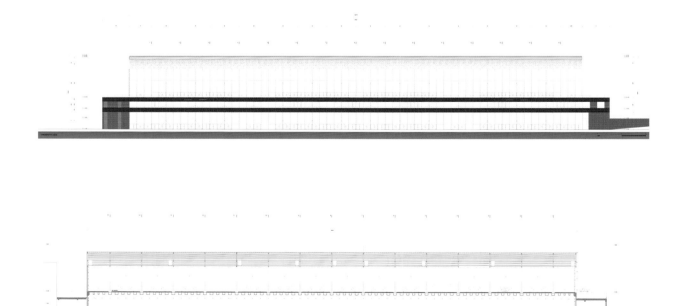

Elevation
立面图

SKIN ANALYSIS 表皮分析

LIGHT SKIN

Large mirror-like roof and glass façade from the enclosed structure. The reflective roof and open-view glass make the entire building does not exist, coexisting with the environment naturally.

轻表皮

镜面般的大型屋顶以及玻璃立面是建筑的主要围合结构,能倒映出蓝天白云的屋顶和视野无阻碍的玻璃让整栋建筑就如同不存在一般,和环境自然共存。

The first mirror is like a kaleidoscope whose geometric nature has been perturbed by randomness. The second, with its parallel rhythm, stretches and deforms images like a car body or bumper.

Beneath the first mirror, light is provided by fixtures that are "plugged into" the floor. They never touch the ceiling. The second mirror is lit by floodlights hidden in the hollows of its waves or behind one-way mirrors. The two systems create a silvery, glistening effect like sunlight on the sea. From the quayside, the exhibitions create a double spectacle of multicoloured, moving reflections.

The restaurant stretches along the limit between the port and the exhibit halls and enjoys a view, depending on which direction one looks in, of the serenely moored boats in the port, or of the animation in the exhibit hall.

This is a story about questioning, about the destruction of material, a story of high-class optical games, of delicious torments for the eyes on the scale of a city.

These sensations are put in the service of making exhibit halls into attractive, prestigious places, for they, too, should love a show.

第一间大厅的镜面屋顶就像是一个万花筒,其几何特性被不规则的形状打乱;第二间大厅的屋顶具有平行的线条,就像是像是车身或保险杠、可延伸和变形。

第一间大厅的镜面屋顶下的照明不会接触到天花板,而是通过嵌入地板的灯具实现。第二间大厅的镜面屋顶则采用隐藏在波浪的凹陷处或单向镜后的泛光灯作照明。两种照明系统打造出银色闪光的效果,如同海面上反射的阳光。从码头上看,展厅创造了一个多彩、动感的双重景象。

餐厅位于港口和展厅之间的空隙中,其周边景色会随着观赏的角度不同而有所不同,有时会看到安静地停泊在港口的船只,有时则会看到充满动感的展览大厅。

本案的设计与设计师的探索精神、材料的毁灭与重生、高端的视觉效果,以及吸引眼球的城市规模有着密切的联系。

它们使该建筑成为热爱展览的观众中最具吸引力和声望的场所。

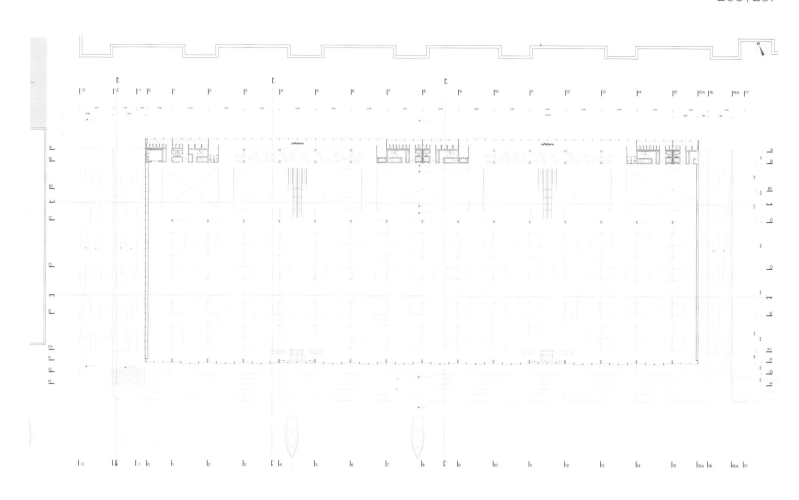

Floor Plan 1
楼层平面图一

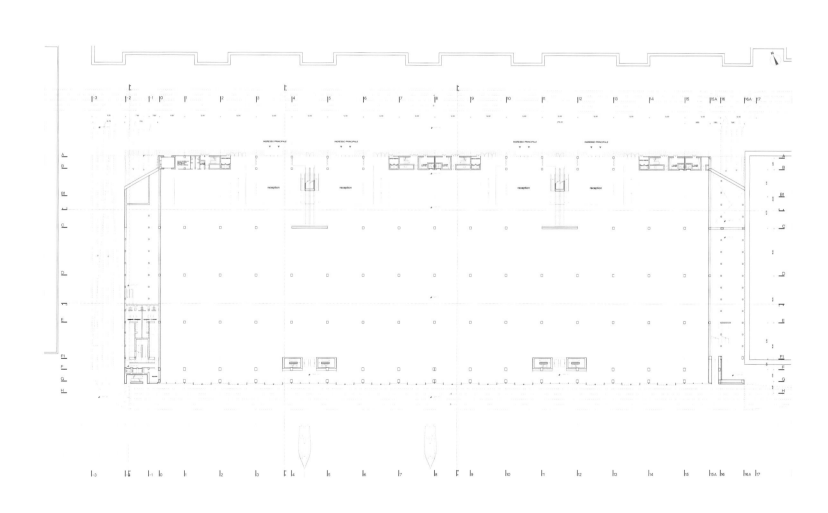

Floor Plan 2
楼层平面图二

FERRARI WORLD ABU DHABI

阿布扎比酋长国法拉利主题公园

Architect: Benoy
Location: Abu Dhabi
Site Area: 236,000 m²

设计公司：贝诺建筑事务所
地点：阿布扎比酋长国
占地面积：236 000 m²

STRUCTURE AND MATERIAL 结构与材料

STRUCTURE
"Qround Hugging" Structure

MATERIAL
Insulated Metal, Glass

结构
"伏地"结构

材料
绝缘金属、玻璃

At the centre of the landmark Yas Island megascheme in Abu Dhabi, sits the world's first vvv Theme Park a thrilling brand experience like no other, a multi-sensory celebration of a design icon.

Ferrari World Abu Dhabi's location, scale and purpose combined presented enormous architectural challenges. In response, Benoy delivered a revolutionary design solution. The end result is an iconic landmark leisure destination that reflects both the integrity of the Ferrari brand and the ambitions of Abu Dhabi.

Opened in December 2010, Ferrari World Abu Dhabi is the world's largest indoor theme park.

Benoy's vision was to create a building that would reflect the highly recognizable sinuous form of Ferrari, directly inspired by the classic double curve profile of the Ferrari GT chassis.

世界首个法拉利主题公园坐落于阿布扎比酋长国亚斯岛中心，将为广大法拉利爱好者们带来前所未有的强烈多感官体验。

如何将法拉利主题公园的位置、规模和用途巧妙地结合在一起成为本案最大挑战。设计师对此也提出了革命性的设计方案，最终打造了一座休闲娱乐的地标建筑，同时又反映出法拉利品牌的完整性和阿布扎比的雄心壮志。

法拉利主题公园于2010年12月开业，将成为世界上最大的室内主题公园。

设计师希望该建筑能表现出法拉利经典的曲线造型，其设计灵感直接来源于法拉利GT跑车底盘的经典侧面双曲线造型。

Site Plan
总平面图

SHAPE ANALYSIS 造型分析

PICTOGRAPHIC SHAPE
The red giant triangle star like a UFO landed here, forming a distinct contrast with the surrounding buildings and environment to highlight the project's eye-capturing uniqueness.

象形造型
巨型的鲜红三角星就像是一处 UFO 降落点，与周围的建筑和环境都形成鲜明对比，极大地突出了建筑的标新立异，极为夺目。

SKIN ANALYSIS 表皮分析

ENERGY SKIN
The metal skin roof is highly insulated and the main façades utilize efficient glass to reduce thermal loads and glare.

节能表皮
屋顶材料选用高度绝缘金属，主立面使用不同的玻璃以减少热负荷和眩光感。

The building was conceived as a simple "Ground hugging" structure: a red sand dune. A three pointed star in plan with extensive "Tri-from" claws to cradle outdoor attractions, the 3D nature of the building was derived from the sinuous double curve of the classic Ferrari body shell. The double curve was proportionally applied in elevation to set the structure's length (700 m) and height (45 m).

At the centre, the roof dips and gathers itself into the ground in the form of a crystal glazed lit funnel, creating the perfect location for one of the most exhilarating rides: the 60 m high "G-Force Tower".

Ferrari World Abu Dhabi houses more than 20 high octane attractions under a geometrically compelling Spaceframe structure. With state-of-the-art simulators allowing guests to experience the thrill of racing a Ferrari and the world's fastest roller coaster reaching speeds in excess of 200kph, Ferrari World Abu Dhabi is an awe-inspiring experience.

建筑可视为一个简单的"伏地"结构：一座红色的沙丘。一个大型的三角星在三个方向各伸出两条"触手"，成为一个抢眼的户外景点。建筑的立体感来源于法拉利的经典双曲线型车身。为了使这个长 700 m、高 45 m 的结构能很好的应用在立面造型上，设计师对原来的汽车双曲线型进行了相应的比例改造。

屋顶在中心往内凹进，以水晶釉面的漏斗结构与地面连接，为公园中最刺激的游客设施之一——60 m 重力加速体验装置的提供了完美的装置地。

此外，在这个夺目的几何结构下还隐藏着法拉利主题公园 20 多个超刺激的空中游乐设施。最先进的是那个能让游客体验刺激的法拉利赛车或是世界上最快的过山车的模拟设施，其速度超过 200 km/h。法拉利主题公园将是一个刺激与惊险并存的娱乐王国。

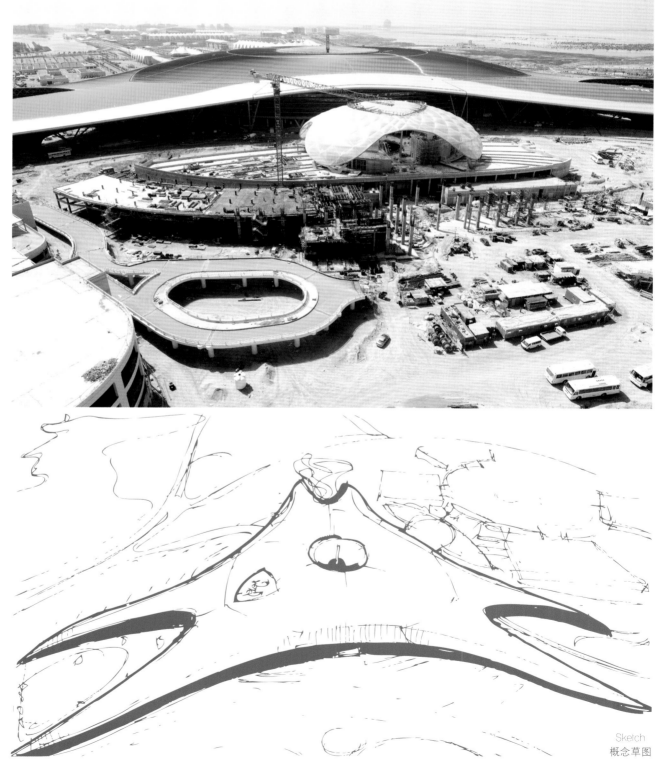

Sketch
概念草图

Triose 商业综合楼

Architect: Sanjay Puri Architects
Client: Shree Vinayak Hospitality Pvt Ltd
Location: Lonavala, India
Built Area: 3,000 m²
Photography: Sanjay Puri Architects

设计公司：Sanjay Puri 建筑事务所
客户：Shree Vinayak Hospitality Pvt Ltd
地点：印度洛纳瓦拉
建筑面积：3 000 m²
摄影：Sanjay Puri 建筑事务所

STRUCTURE AND MATERIAL 结构与材料

STRUCTURE
Reinforced Concrete Structure
MATERIAL
Concrete, Glass

结构
钢筋混凝土结构
材料
混凝土、玻璃

SITE PLAN
1- OFFICINA VIDRE NEGRE
2- GUEST HOUSE
3- GUARDIAN HOUSE

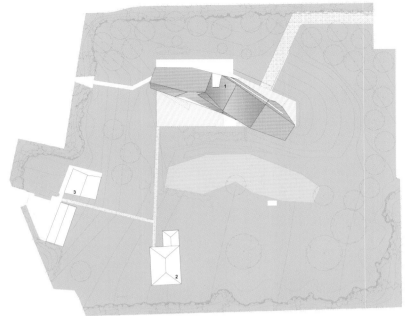

Master Plan
总规划图

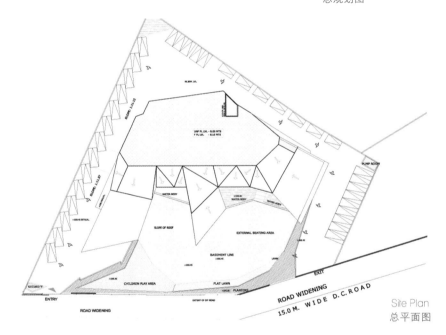

Site Plan
总平面图

Angled spaces projected towards different directions encapsulated in an organically folded concrete skin, create a two level building on this site in Lonavala, housing a few retail shops, a food court, two restaurants, a large bar and an entertainment gaming area.

The entire frontage of the site along the main road overlooks large trees and a riverbed and hills. The axis of the building changes constantly from one side to the other allowing each space within to look out towards different views of the surrounding landscapes.

The concrete folded skin that forms most of the building creates large open frames towards the external views and the plans of the building too open out towards these large frames accentuating the beautiful natural surroundings to the inner spaces. The building is comprised of three volumes which emanate from a central circulation spine that interconnects them.

The building is created sculpturally from within & externally and is a unique manifestation of abstracted volumes that are fluid in the interior and perceived as a dramatic juxtaposition of trapezoidal volumes on the site.

本案坐落在风景如画的洛纳瓦拉小镇，采用有机折叠混凝土作为表皮，塑造了多个朝不同方向延伸的菱角，构建出一栋两层楼高的建筑。建筑内部设置了多个零售商店、两间餐厅、一个美食广场、一个大型酒吧，以及一个休闲娱乐场所。

本案正前方临近主干道，站在这里便可俯瞰大片树林、湛蓝的河床以及连绵起伏的丘陵。由于建筑轴线不断变换，使得每个空间里都有独特的观景角度，能欣赏不一样的自然景观。

折叠起伏的混凝土构成了建筑的大部分体量，也创造了一个宽敞而开放的框架，既能方便欣赏户外风景，又能衬托出周边景观的美态。在造型上，本案由三个庞大的开放式悬空体量组合而成。它们从一个中央脊柱出发，向四周辐射延伸，彼此以不同的方式穿插连通。

室内与室外的设计均体现出雕塑感。在室内渲染雕塑质感，在室外则通过一种不规则的梯形体量，最终将抽象的体量与具体的结构戏剧性地结合起来。

SHAPE ANALYSIS 造型分析

DIGITAL SHAPE
Geometric modeling changing continuously is based topology principles that subverts the traditional architectural images.

数字造型
本案利用拓扑学原理设计出连续变化的几何造型，这种设计完全颠覆了传统建筑的形象。

SKIN ANALYSIS 表皮分析

HEAVY SKIN
Concrete skin endows sculptural sense to the building with its brown color to be a transtional between the building and the environment.

重表皮
本案采用混凝土作表皮，赋予了建筑硬朗的雕塑质感。而土黄色的外皮漆料则让雕塑感较重的建筑与周边环境过渡自然。

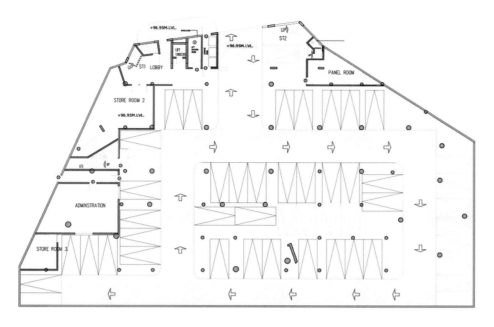

Basement
地下室

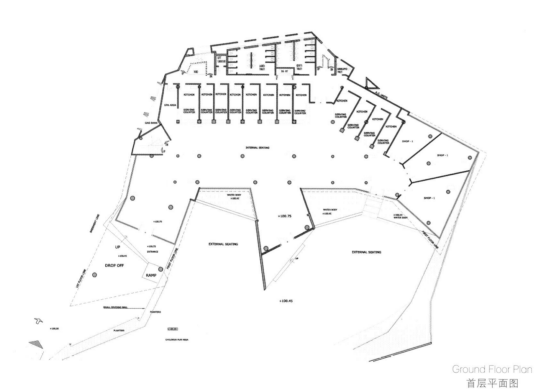

Ground Floor Plan
首层平面图

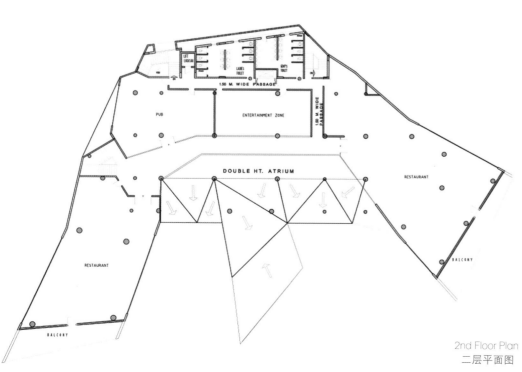

2nd Floor Plan
二层平面图

Markthof 综合大楼

Architect: Pi de Bruijn, de Architekten Cie.
Client: Markthof c.v., Amsterdam
Location: Hoofddorp, The Netherlands
Built Area: 28,000 m²

设计公司：Pi de Bruijn、de Architekten Cie.
客户：Markthof c.v.、Amsterdam
地点：荷兰霍夫多普
建筑面积：28 000 m²

STRUCTURE AND MATERIAL 结构与材料

STRUCTURE
Reinforced Concrete Structure

MATERIAL
Steel, Glass, Brickwork

结构
钢筋混凝土结构

材料
钢材、玻璃、砖

Markthof is a newly built complex, replacing an existing shopping mall from the 1980s. It boasts a functional mix of shops, offices, parking spaces and apartments. Like a conjuring trick, it has transformed an existing, introverted shopping centre into a versatile urban unit. As a unique feature of this location, the urban block is built around a monumental tree. Its unique position on the Dik Tromplein makes the Markthof the cornerstone at the heart of this urban centre.

Markthof's layout creates an appealing interplay between the complex's different functions. The shopping block is accessible on three sides, injecting life into the adjacent streets. The commercial functions are on the ground and first floors. All give on to the outer façade and thus contribute to a varied street-scape. The car park is a special feature, located above the shops and hidden perfectly behind their façades. Above this level are the rental office space and residential functions.

The varying surroundings, ranging from narrow shopping street to open square, dictate Markthof's own strong, contrasting architecture.

The long walls of the shop fronts are divided into vertical brick strips, of different widths and heights, in order to preserve the individual identities of the different shops. By creating a living space around a rooftop garden/square, amid all the facilities of the centre, they instil a special sense of privacy. The dark colour of the brick and a strong, horizontal subdivision of the façade underline that the apartments are a private sphere, in contrast to the public sphere of the shops.

The glazed box, which houses the rental office spaces on the corner of the Kruisweg and Dik Tromplein, catches the eye. Through its glass curtain wall, this "pavilion" displays playfully curved structural steelwork. The interplay of arcs is inspired by the fruit of the wingnut tree.

This striking element has become iconic, representing not only Markthof, but the whole renewed centre of Hoofddorp.

Markthof 是一座新建的综合大楼，取代了建于上世纪八十年代的、原有的购物中心。大楼内包含一个多功能商店、办公室、停车场和公寓。它戏剧般地将现有的循规蹈矩式的购物中心转变成为一个多功能的城市单元。这个地区最独特的地方在于，整个街区围绕着一棵具有纪念意义的树而建，而这使 Markthof 成为 Dik Tromplein 城市中心的标志建筑。

Markthof 综合大楼的布局为建筑内的不同功能区间创造了一种联系，使其相互联通、相互影响。大楼内的购物区可以通过建筑的三个侧面进入，充分融入邻近的街区。商业区分布在地面层和一层。所有的空间都成为外立面的一部分，为人们创造了风情万种的街景。停车场位于购物区上方，被立面完美地隐藏在建筑中。除了这些，大楼的高层部分则被用作出租办公区和住宅区。

从狭窄的购物街道到开阔的广场，不同的环境造就了 Markthof 综合大楼鲜明的建筑特征。

店面的长墙被划分成多个不同的宽度和长度的垂直砖带，以保持不同商店的独立性。通过围绕着屋顶设置一个居住空间，设计师创造了一种特别的隐私感。深色的砖块配上明显的横向分割设计，将大楼的公寓部分独立出来，与商店等公共场所形成了鲜明的对比。

大楼上方的"玻璃盒"是办公区，盘踞在 Kruisweg 和 Dik Tromplein 之间，格外引人瞩目。穿过它的玻璃幕墙，人们可以看到那充满趣味性的、弯曲的钢结构。弧光灯的使用主要是受了那棵螺母般的大树的果实所启发。

这些亮眼的元素俨然成为一种标志，不仅代表了 Markthof，更标志着 Hoofddorp 中心的重生。

SHAPE ANALYSIS 造型分析

OPEN SHAPE

The building layout creates a link applied between different functions of the building. Independent stores, secret private space and a different style of street view, all of these give the best interpretation for the uniqueness of the building.

开放式造型

建筑的布局在其不同功能之间创造了一种联系。独立的商店、隐秘的私人空间，以及不同风情的街景，都为建筑的独特性做了最好的诠释。

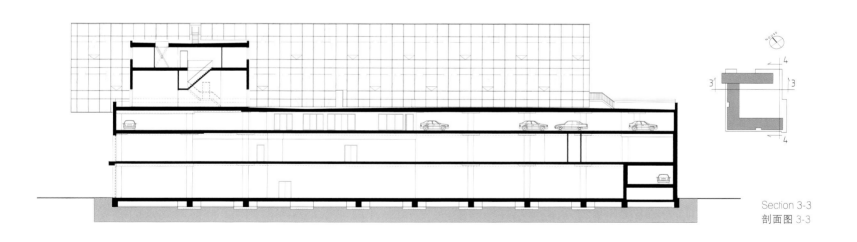

Section 3-3
剖面图 3-3

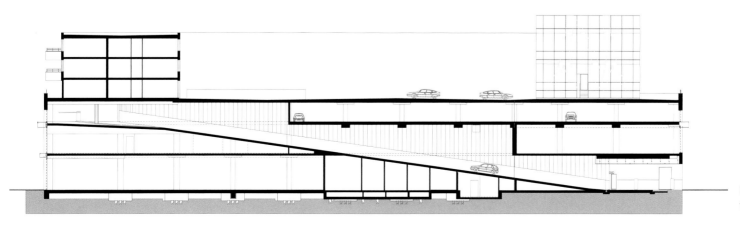

Section 4-4
剖面图 4-4

5th Floor Plan
五层平面图

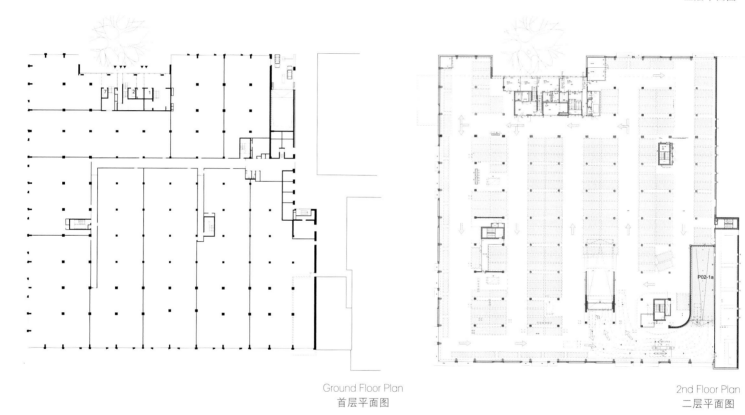

Ground Floor Plan
首层平面图

2nd Floor Plan
二层平面图

SKIN ANALYSIS 表皮分析

TRANSPARENT SKIN
The office spaces on the corner of the Kruisweg and Dik Tromplein has luxuriant glass surface, and through its glass curtain wall one can see structural steelwork.

透明表皮
位于 Kruisweg 和 Dik Tromplein 拐角处的办公区拥有华丽的玻璃表皮，透穿其玻璃幕墙，人们可以看到整个建筑的钢结构。

HOTEL
酒店

SOFITEL VIENNA STEPHANSDOM
索菲特维也纳斯蒂芬斯顿酒店

Architect: Ateliers Jean Nouvel
Artistic Conception: Pipilotti Rist
Client: UNIQA Praterstrasse Projekterrichtungs GmbH
Location: Vienna, Austria
Built Area: 53,000 m²
Photography: Philippe Ruault, Roland Halbe

设计公司：让·努维尔事务所
艺术概念设计：Pipilotti Rist
客户：UNIQA Praterstrasse Projekterrichtungs GmbH
地点：奥地利维也纳
建筑面积：53 000 m²
摄影：Philippe Ruault、Roland Halbe

STRUCTURE AND MATERIAL 结构与材料

STRUCTURE
Reinforced Concrete Structure

MATERIAL
Granited Glass, Concrete

结构
钢筋混凝土结构

材料
花岗岩玻璃、混凝土

Architecture is the art of taming constraints, of poetizing contradictions, of looking differently at common and trivial things in order to reveal their singularity. Architecture is an opportunity, in a city marked by history, to continue games begun by others years or centuries ago. It is a clever game of chance and intention; an occasion to modify, to deepen, or to change the meaning of a context. Architecture is about making apparitions.

建筑是一门揭示独特的艺术。设计师克服了限制条件，将矛盾诗意化，从琐碎平凡的事物中看出特点，并加以应用。在历史悠久的城市里，建筑的存在就是为了延续几年前或几个世纪以前就已经开始的游戏。这是一场靠技巧和意向取胜且充满机遇的游戏。建筑可能是一次改良、一次深化或一次改变环境意义的机会。总而言之，建筑就是建造奇迹。

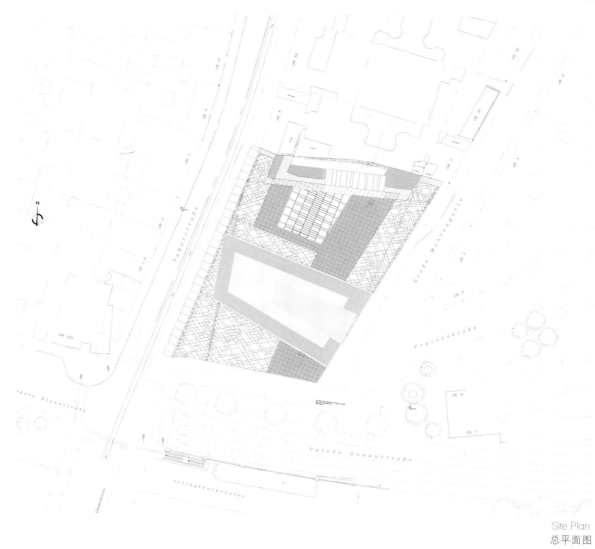

Site Plan
总平面图

Section 1　　　　　　　　　　　　　　　　Section 2
剖面图一　　　　　　　　　　　　　　　　剖面图二

SHAPE ANALYSIS 造型分析

OPEN SHAPE
The unique shape is characterized by a medium "long mouth" and a roof canopy with color patterns, which seems to attract people into a closer look.

开放式造型
本案造型极为独特,建筑中部被"挖"出一道形似三角体的"口子",而且这道"口子"的天花板与顶楼的天篷均能照映出五彩斑斓的图案,吸引着人们走近细看。

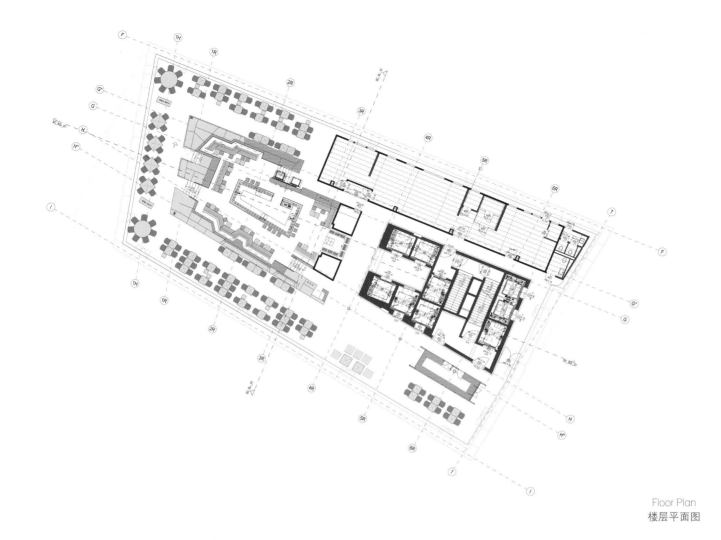

Floor Plan
楼层平面图

So just imagine that starting with these curious constructible prisms, their planes begin to slide; intersections are created; one plane begins to tilt under the magnetic deviance of HH while another decides to light the city from a ceiling made of furtive images. Imagine that the other planes begin to vibrate with a thousand lines of variable orientation and reflectivity, that gray sometimes melts into gray squares on a gray background. It is not surprising then to find that the oblique plan of the roof becomes hatched, weaving a tight, random pattern of parallelograms and lozenges. At the limit between building and sky there is another, flat plane that reveals the appearance-disappearance of changing faces, an evocation of the multiple faces forever linked to the depth of imagery born of this city.

所以试想下，从这些新奇独特的棱镜开始，被棱镜覆盖的平面开始滑动并构筑建筑物的各个立面；它们从一个交叉点开始，一个平面在正负磁力偏差装置的作用下倾斜，另一个平面则借由天花板上隐蔽的图案点亮了城市的风景。试想一下，那些灰暗的平面被成千上万条不断改变映射方向的光线所照亮，有时，甚至那些暗淡的光线渐渐消失在灰暗立面中的灰色方块里，这是多么神奇的景象。所以，当人们从远方看到倾斜的屋顶平面上那些平行四边形或菱形图案时而黯淡熄灭、时而交织紧凑、时而随意搭配的样式时，也不会感到惊讶。在建筑与天空的相交线上，那里有另一个与地平线平行的平面也在展示瞬息转换、忽明忽暗的图案。这就是设计师的意图，将瞬息变幻的平面与城市风景联系起来，并以此唤起人们对这栋建筑的回忆。

SKIN ANALYSIS 表皮分析

LIGHT SKIN

The building is covered by glazed curtain wall. The planes to the North take the form of granited glass for transparence; that the planes to the West cloak themselves in variations of black to display their shadows.

轻表皮

本案整体被玻璃幕墙覆盖着。北立面使用花岗岩玻璃营造透明质感；西立面则采用各种黑色元素表现阴影感。

Floor Plan
楼层平面图

THE YAS HOTEL
Yas 酒店

Architect: Asymptote Architecture
Client: Aldar Properties PJSC
Location: Abu Dhabi, United Arab Emirates
Built Area: 85,000 m²

设计公司：渐近线建筑事务所
客户：Aldar Properties PJSC
地点：阿联酋阿布扎比市
建筑面积：85 000 m²

STRUCTURE AND MATERIAL 结构与材料

STRUCTURE
Steel Structure
MATERIAL
Steel, Diamond-shaped Glass Panels

结构
钢结构
材料
钢筋、钻石玻璃板

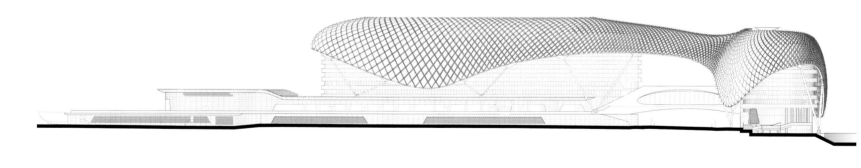

SOUTH ELEVATION 0 10 20 40
SCALE IN METERS

South Elevation
南立面图

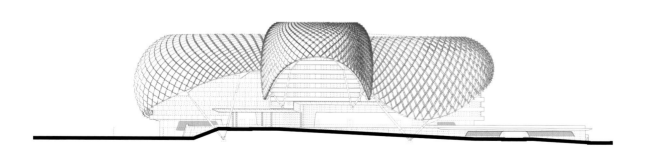

WEST ELEVATION 0 10 20 40
SCALE IN METERS

West Elevation
西立面图

SHAPE ANALYSIS 造型分析

PICTORAGPHIC SHAPE

The surrounding environment of the project is the ocean and desert. The Grid - Shell hull covers the main building and present the visual effect like gems.

象形造型

蔚蓝的大海与广袤的沙漠赋予本案非一般的区位环境。形似贝壳外壳的网状结构覆盖着主建筑，使主建筑如宝石般被珍藏在网状贝壳内，呈现出独特的视觉效果

The Yas Hotel, a 500-room, 85,000-square-meter complex, is one of the main architectural features of the ambitious 36-billion-dollar Yas Marina development and accompanying Formula 1 raceway circuit in Abu Dhabi, UAE.

Of architectural and engineering significance is the main feature of the project's design, a 217-meter expanse of sweeping, curvilinear forms constructed of steel and 5,800 pivoting diamond-shaped glass panels. This Grid-Shell component affords the building an architecture comprised of an atmospheric-like veil that contains two hotel towers and a link bridge constructed as a monocoque sculpted steel object passing above the Formula 1 track that makes its way through the building complex.

本案占地 85 000 m²，总套房数 500 间，总投资额高达 360 亿美元，是 Yas Marina 大型发展项目中最重要的特色建筑之一，与阿布扎比市 F1 赛场相毗邻。

建筑工程的意义就是本案的主要建筑特色的所在。它长达 217 m，将整个外立面拥抱在"怀"，富有曲线美的造型采用钢筋和 5 800 个绕轴旋转的钻石玻璃板打造而成。形如贝壳的网状结构轻薄得如大气层一般，笼罩着酒店的两栋主楼，覆盖着一条连接主楼并横跨 F1 赛道的雕刻式钢架桥，让酒店与 F1 赛道形成一个庞大的综合性建筑。

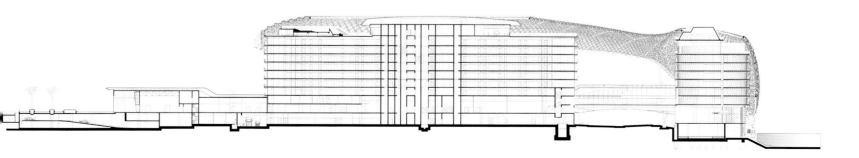

SECTION AA

Section A
剖面图 A

SKIN ANALYSIS 表皮分析

TRANSPARENT SKIN

Building façade is glass curtain wall, and most were hung over by the mesh structure, this structure can reduce the sunshine, at the same time meet the need of daylighting and ventilated. LED lights are installed on the mesh structure, when the dark night comes, it presents a wonderful visual enjoyment.

透明表皮

本案外立面打造玻璃幕墙的效果，大部分建筑被形似贝壳外壳的网状结构笼罩着。该结构在减少光照直射的同时，满足了采光和通风的需要。当夜幕降临时，被安装在网状结构上的LED灯光会呈现出美轮美奂的视觉效果。

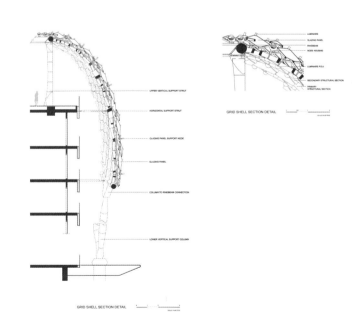

Shell Detail Section
外壳节点剖面图

The Grid-Shell visually connects and fuses the entire complex together while producing optical effects and spectral reflections that play against the surrounding sky, sea and desert landscape. The architecture as a whole "performs" as both an environmentally responsive solution as well as an architecture of spectacle and event. The entire jewel-like composition of the project responds visually and tectonically to its environment to create a distinct and powerful sense of place as well as a breathtaking backdrop to the Formula 1 and other events that the building will celebrate. The Yas Hotel is designed to be a significant landmark destination on Yas Island for Abu Dhabi and the UAE at large.

形如贝壳的网状结构把两座主楼合二为一，呈现出视觉上的统一。网格上的LED灯光在夜幕下所产生的光学效应和光谱反射，形成可与漫天星空、茫茫海洋和沙漠风光相媲美的美景。本案就像一个全能的舞台，它既展示成功的环保案例，也呈现奇特的景观和奢华的设计。本案宝石般的造型，是栖息自然而诞生的视觉盛宴，是巧用地形而造就的强大构造。让人惊喜的是，它还背靠着F1赛场，适合举办各类大型庆典。Yas酒店的设计对于阿布扎比市的亚斯岛，甚至对于整个阿联酋来说，都具有重大的里程碑式意义。

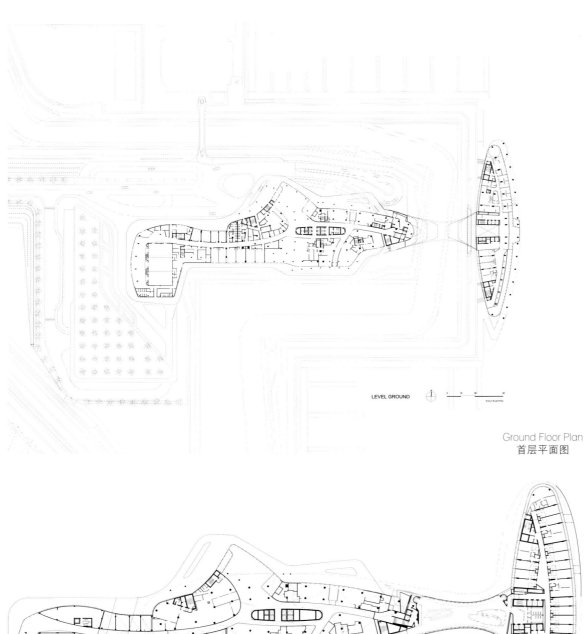

Ground Floor Plan
首层平面图

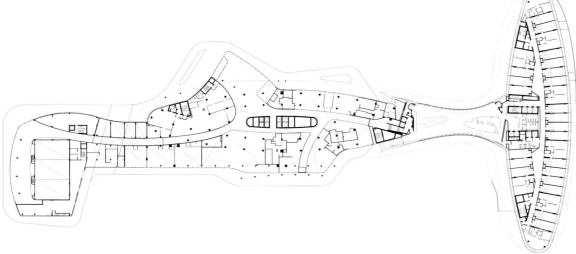

Mezzanine Floor
阁楼平面图

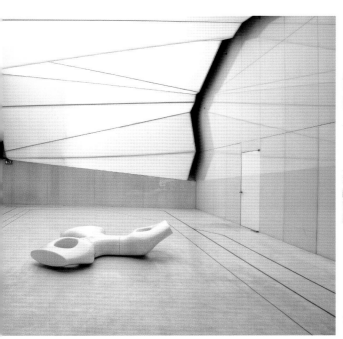

OMNI DALLAS HOTEL
欧姆尼达拉斯酒店

Architect: 5G Studio
Architect of Record: BOKA Powell
Location: Dallas, Texas, U.S.A.
Site Area: 33,884.33 m²
Photography: 5G Studio

设计公司：5G Studio
施工单位：BOKA Powell
地点：美国德克萨斯州达拉斯市
占地面积：33 884.33 m²
摄影：5G 工作室

STRUCTURE AND MATERIAL 结构与材料

STRUCTURE
Reinforced Concrete Structure
MATERIAL
Crème Limestone, Glass Curtainwall

结构
钢筋混凝土结构
材料
乳白石灰石、玻璃幕墙

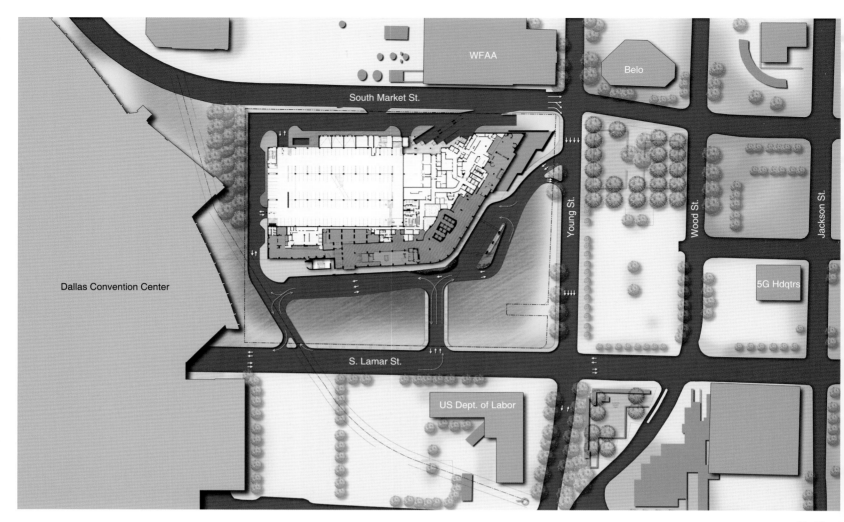

Site Plan
总平面图

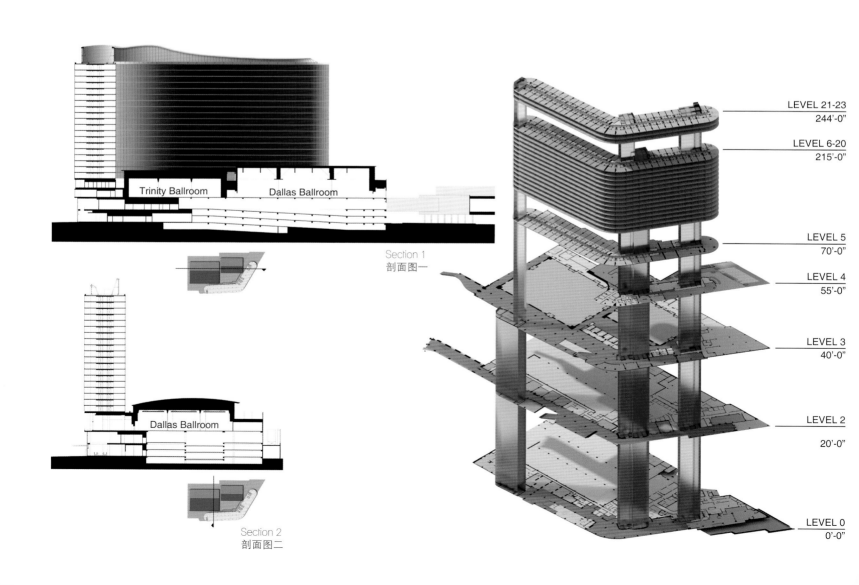

Section 1
剖面图一

Section 2
剖面图二

SHAPE ANALYSIS 造型分析

OPEN SHAPE

The design of the building takes more consideration on the interaction of the local culture and economy, and the translucent square entrance will link the building and street together.

开放式造型

本案以促进当地文化与经济的交流为理念进行规划设计。设计师匠心独运,为广场打造了半透明的入口,远远望去,酒店与街道仿若相连,形同一体。

SKIN ANALYSIS 表皮分析

TRANSPARENT SKIN

The selection of materials for the exterior of the building, highly transparent glazing on the north and east façade, allow for passersby to catch glimpses of activity within the space. The rounded ends of the tower provide end suites with an expansive 180-degree view of the North and South Dallas skyline.

透明表皮

在本案中,朝向东、北的两个外立面运用了玻璃材质。这种材质透明度极高,路人在经过建筑的北面和东面时,可随意观看酒店内的情况。除此以外,顶楼最高处的圆拱也运用了玻璃。透明圆拱为游客提供了180°的广阔视角,使本案的顶楼成为观赏达拉斯南北地平线的最佳位置。

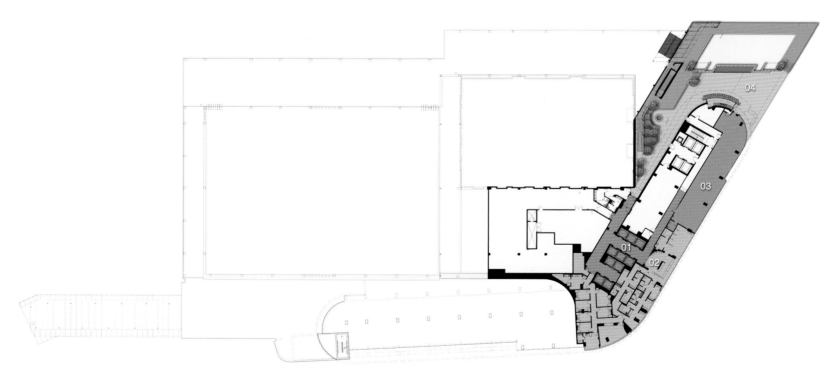

01 Circulation
02 Mokara Spa
03 Fitness Center
04 "Uptown Terrace" Pool Deck

4th Floor Plan
四层平面图

Nestled in the center of downtown Dallas, adjacent to the 2.1 million square foot Dallas Convention Center, the Omni Dallas Convention Center Hotel marks the beginning of a new, revitalized city center. The Omni occupies six of eight acres at the corner of Harwood and Young streets, rising 23 stories and containing 1,016 rooms. The remaining acres on the site will be occupied by dining, retail, and other as yet to be determined services.

The geometry and diagram of the building were structured to activate Lamar Street and create a seamless, integrated interaction between the guests and passersby alike. In addition to interacting with side streets, the building responds to the neighboring architecture of the Convention Center itself. To maximize accessibility to the Convention Center amenities and promote spatial continuity between the two structures, the hotel's public-use spaces are located on the lower level. These include more than 80,000 square feet of meeting space, several restaurants, lounges, and a signature spa. The tower of rooms is perched atop massive columns and a transparent base, presenting the hotel to the city.

While substantial in scale and function, the building form overwhelms neither the Convention Center, nor the neighboring streets. The translucent and spacious entry plaza provides a seamless and unencumbered integration between street and structure, inviting guests in without overpowering or intimidating. Additionally, the design accounts for multiple restrictions and opportunities: limiting site constraints, a tight construction schedule, and LEED Silver certification.

The Dallas Convention Center Hotel is pursuing LEED Silver certification from the USGBC. The site is located within a ½ mile of 2 DART stations, allowing easy access to public transportation for employees and guests, resulting in a lower carbon footprint for travel to and from the hotel. The hotel will include preferred parking for low-emitting and fuel-efficient vehicles; bicycle racks will be present on site as well. In order to reduce the onsite heat island effect, the roofing materials and site paving will have high solar reflectance index values.

本案坐落在达拉斯的闹市中心，毗邻占地 195 096.39 m² 的达拉斯会议中心，本着发展城市活力的宗旨进行规划创建。它位于哈姆德街和新街的交界处，总面积达 33 884.33 m²，高 23 层，共 1 016 间客房，其余空间将开设餐厅、商店、以及其他尚未确定的服务设施。

本案的几何形态和图案造型以方便往来的顾客与行走在马尔街的人们交互相见为目的。那么，如何加强人与人之间无隔阂的沟通互动呢？本案将公共活动区设在底层，使活动区直接与会议中心相连，表现出两栋建筑的空间连续性。活动区占地 7 432.24 m²，涵盖了会议厅、餐厅、休息室和特色水疗设施等。酒店中的客房位于同一栋塔楼中，由几根粗大的柱子撑起并倚靠在半透明底座上，能观赏到最美的城市风景。

在造型设计和职能组织上，本案既不会盖过会议中心的气派，也不会给社区街道造成影响。底层的广场入口宽阔精致，半透明的材质使得酒店与街道衔接自然，典雅活泼的设计更容易吸引顾客进场消费。此外，本案的设计方案还将多种因素考虑在内，包括有限的场地、紧凑的工期、以及 LEED 银级认证标准等。

本案为成功申请美国绿色建筑委员会颁发的低能量电子衍射银级认证证书，设计了三种低排节能方案。其一，酒店鼓励员工和客人乘坐公共交通工具往返出行。酒店距离 DART 车站只有 805 m 的路程，乘坐交通工具有助于降低汽车尾气中的碳排放量。其二，优先给予低排环保和高效节油汽车停车位，也提供自行车停放架，倡导自行车出游。其三，为减缓热岛效应，屋顶和场地材料优先选用能高效反射太阳热能的材质。

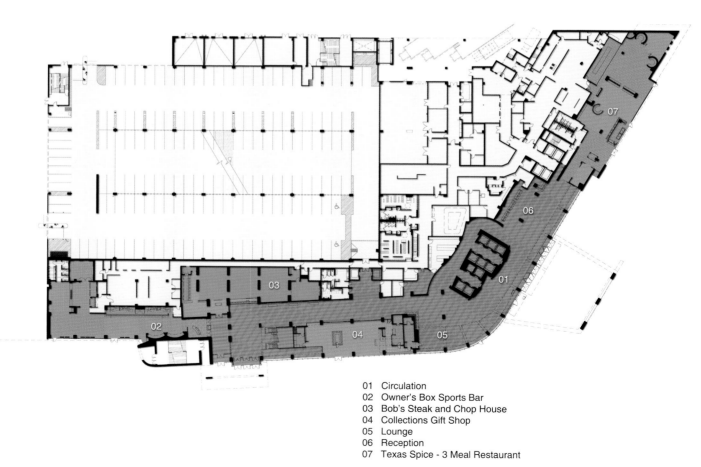

01 Circulation
02 Owner's Box Sports Bar
03 Bob's Steak and Chop House
04 Collections Gift Shop
05 Lounge
06 Reception
07 Texas Spice - 3 Meal Restaurant

Ground Floor Plan
首层平面图

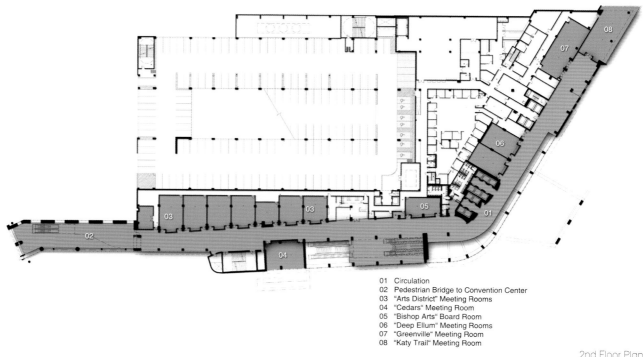

01 Circulation
02 Pedestrian Bridge to Convention Center
03 "Arts District" Meeting Rooms
04 "Cedars" Meeting Room
05 "Bishop Arts" Board Room
06 "Deep Ellum" Meeting Rooms
07 "Greenville" Meeting Room
08 "Katy Trail" Meeting Room

2nd Floor Plan
二层平面图

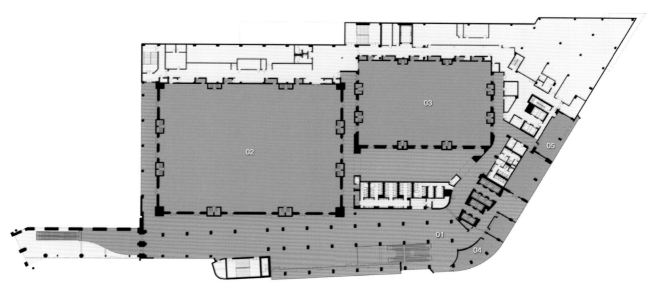

01 Circulation
02 "Dallas" Ballroom
03 "Trinity" Ballroom
04 "Southside" Meeting Rooms
05 "Fair Park" Meeting Rooms

3rd Floor Plan
三层平面图

DREAM DOWNTOWN HOTEL 梦想中心酒店

Architect: Handel Architects
Client: Hampshire Hotels & Resorts, Vikram Chatwal Hotels
Location: New York City, U.S.A.
Site Area: 2,350.44 m²
Photography: Adrian Wilson, Bruce Damonte

设计公司：Handel Architects
客户：新罕布什尔州酒店及度假村、Vikram Chatwal 酒店
地点：美国纽约市
占地面积：2 350.44 m²
摄影：Adrian Wilson、Bruce Damonte

STRUCTURE AND MATERIAL 结构与材料

STRUCTURE
Frame Structure
MATERIAL
Stainless Steel, Glass

结构
框架结构
材料
不锈钢、玻璃

334 335

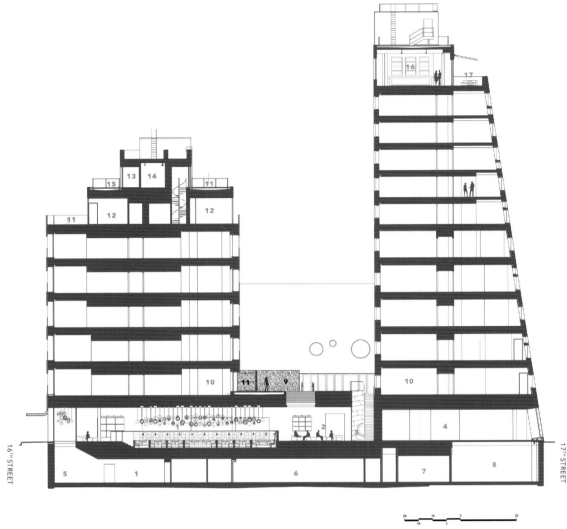

1	RESTAURANT	10	GUESTROOMS
2	LOBBY / LOUNGE	11	GUESTROOM TERRACES
3	GARDEN OPEN TO ABOVE	12	GUESTROOM PENTHOUSE SUITE
4	EVENT / EXHIBITION	13	GREENHOUSE
5	OPEN TO ABOVE	14	MECHANICAL
6	RESTAURANT KITCHEN	15	TERRACE
7	HOTEL BACK OF HOUSE	16	ROOF LOUNGE
8	LOUNGE	17	ROOF LOUNGE TERRACE
9	POOL / BEACH		

Section
剖面图

SHAPE ANALYSIS 造型分析

PICTOGRAPHIC SHAPE

Frontage elevation adopts the slant stainless steel metope, like a mirror reflects the blue sky, white cloud and the sun.

象形造型

临街的立面采用倾斜的不锈钢墙面设计，像一面镜子反射着蓝天、白云和太阳。

Dream Downtown Hotel is a 17,094 m² boutique hotel in the Chelsea neighborhood of New York City. The 12-story building includes 316 guestrooms, two restaurants, rooftop and VIP lounges, outdoor pool and pool bar, a gym, event space, and ground floor retail. Dream sits on a though-block site, fronting both 16th and 17th Streets. Along the 17th Street exposure, the sloped façade was clad in stainless steel tiles, which were placed in a running bond pattern like the original mosaic tiles of Ledner's Union building. New porthole windows were added, one of the same dimension as the original and one half the size, loosening the rigid grid of the previous design, while creating a new façade of controlled chaos and verve. The tiles reflect the sky, sun, and moon, and when the light hits the façade perfectly, the stainless steel disintegrates and the circular windows appear to float like bubbles. The orthogonal panels fold at the corners, continuing the slope and generating a contrasting effect to the window pattern of the north façade.

梦想中心酒店是一个面积为 17 094 m² 的精品酒店，位于纽约附近的切尔西。这座 12 层高的建筑物拥有 316 间客房、两间饭店、顶层贵宾室、户外泳池、池畔吧、健身中心，以及地面零售商店。建筑物同时面朝着 16 和 17 街，其倾斜的立面被不锈钢墙砖覆盖着，像经典的 Ledner's Union building 镶嵌地砖一样，采用顺砖砌合的方式进行铺砌。建筑物添加了新的舷窗，一些与原来的窗子尺寸相同，另一些则只有原窗的一半大小，改变了一贯严格的设计风格，形成了一种有序且富有活力的立面。不锈钢墙砖反射着蓝天、太阳和月光，当阳光极佳地照射在立面上时，不锈钢面砖将散射着反光，使圆形的舷窗看起来就像一个个漂浮的泡沫。直角面板在转角处折合延续着倾斜的形态、生成与北立面窗户的对比效果。

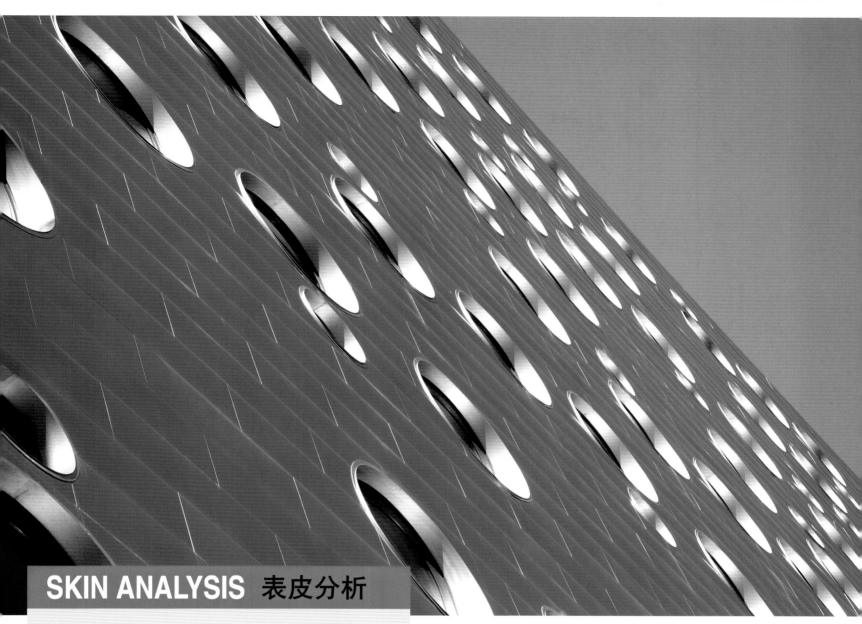

SKIN ANALYSIS 表皮分析

LIGHT SKIN

The skin of building is composed by stainless steel brick, hole wear metal panel and circular windows, the chosen of material and window give a better reflect to the guests.

轻表皮

建筑的表皮由不锈钢砖、孔穿金属面板和圆形窗组成，在材料和窗框的形态选择上都给客人美好的印象。

The 16th Street side of the building, previously a blank façade when the building served as an annex, was given new life. The skin is constructed of two perforated stainless steel layers, its top sheet of holes a replication of the 17th Street punched-window design and the inner sheet a regular perforation pattern. The outer rain screen is punctured with porthole-shaped Juliet balconies for the guestrooms and peels up at the ground level to form the hotel canopy and reveal the hotel entrance. The original through block building offered limited possibilities for natural light. Four floors were removed from the center of the building, which created a new pool terrace and beach along with new windows and balconies for guestrooms. The glass bottom pool allows guests in the lobby glimpses through the water to the outside (and vice versa) connecting the spaces in an ethereal way. Light wells framed in teak between the lobby, pool and lower level levels allow the space to flow. Two hundred hand blown glass globes float through the lobby and congregate over The Marble Lane restaurant filling the space with a magical light cloud. Fixtures and furnishings were custom designed for the public spaces and guestrooms to complement the exterior design and to continue the limitless feeling of space throughout the guest experience.

建筑物临 16 街的一面以前作为附楼时立面被涂成了黑色，现在设计师对其重新设计，使其变成了双层带排孔的面板，面板外层的孔仿照临 17 街立面的冲压窗设计，面板内层则是普通的徘孔设计。此外，由于旧的积木式建筑自然采光功能很差，因此，设计师从旧建筑物中心拆除了四层，在空出的平台中为客房新修了窗子和阳台游泳池。泳池的玻璃池底使大堂的客人能透过水瞥见与其相连的外部空间。位于大堂和泳池底更低处之间的柚木框边的采光井还使空间更为明亮。更令人惊喜的是，设计师令 200 个手工吹制的玻璃球漂浮过大堂，聚集在充满奇幻光照的餐厅上方，增加了整个空间的梦幻效果。公共空间和客房的设施均为定制，以配合建筑外观的设计，延续着客人对空间的美好体验。

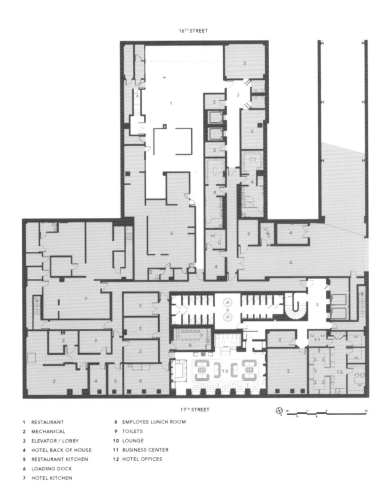

1 RESTAURANT	8 EMPLOYEE LUNCH ROOM	
2 MECHANICAL	9 TOILETS	
3 ELEVATOR / LOBBY	10 LOUNGE	
4 HOTEL BACK OF HOUSE	11 BUSINESS CENTER	
5 RESTAURANT KITCHEN	12 HOTEL OFFICES	
6 LOADING DOCK		
7 HOTEL KITCHEN		

1 ENTRANCE TO BELOW GRADE RESTAURANT	7 CHECK-IN	14 ELEVATOR / LOBBY
2 HOTEL ENTRANCE	8 RAMP	15 EVENT / EXHIBITION
3 SERVICE ENTRY	9 LOBBY LOUNGE	16 EVENT / EXHIBITION ENTRY
4 RESTAURANT	10 FRONT OFFICE	17 TOILETS
5 RESTAURANT BAR	11 KITCHEN	18 MEETING ROOMS
6 LOBBY	12 GARDEN	19 COAT CHECK
	13 HOTEL SHOP	20 LUGGAGE STORAGE

Cellar Plan
地下室平面图

Ground Floor Plan
首层平面图

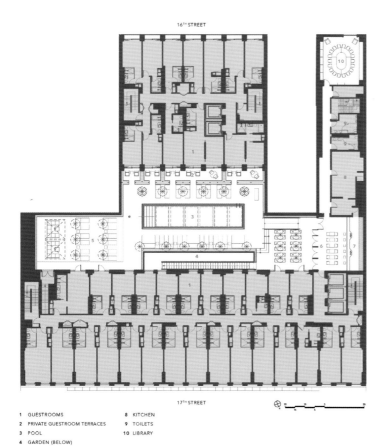

1 GUESTROOMS 8 KITCHEN
2 PRIVATE GUESTROOM TERRACES 9 TOILETS
3 POOL 10 LIBRARY
4 GARDEN (BELOW)
5 BEACH
6 CAFE
7 BAR

2nd Floor Plan
二层平面图

1 GUESTROOMS
2 BALCONIES
3 GYM
4 MECHANICAL

3rd Floor Plan
三层平面图

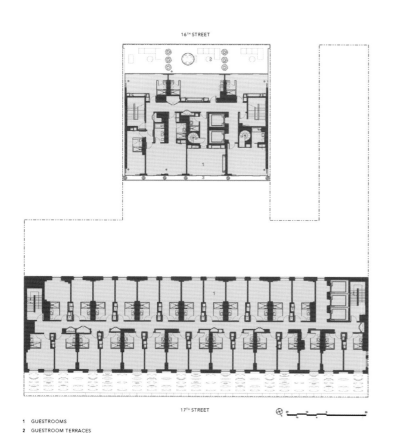

1 GUESTROOMS
2 GUESTROOM TERRACES
3 BALCONIES

7th Floor Plan
七层平面图

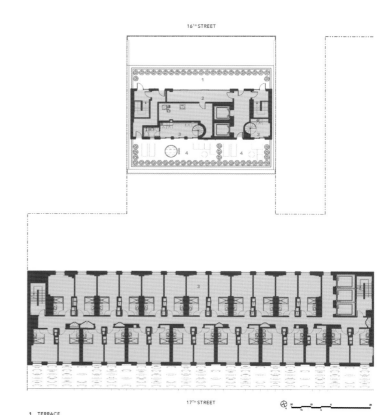

1 TERRACE
2 GREENHOUSE
3 GUESTROOMS
4 PRIVATE GUESTROOM TERRACES

8th Floor Plan
八层平面图

TORRES PORTA FIRA

托雷斯费拉酒店

◆◆◆◆◆◆◆◆◆◆◆◆◆◆

Architect: Toyo Ito & Associates, Architects + Fermín Vázquez - b720 arquitectos
Clients: Hoteles Santos (Hotel Tower), Realia (Office Tower)
Location: Barcelona, Spain
Site Area: 5,755.55 m² (Hotel), 4,801.55 m² (Office)

设计公司：伊东丰雄建筑设计事务所、Architects + Fermín Vázquez - b720 建筑事务所
客户：Hoteles Santos（酒店）、Realia（办公楼）
地点：西班牙巴塞罗那
占地面积：5 755.55 m²（酒店），4 801.55 m²（办公楼）

STRUCTURE AND MATERIAL 结构与材料

STRUCTURE
Reinforced Concrete Structure
MATERIAL
Aluminium Panel, Glass

结构
钢筋混凝土结构
材料
铝板、玻璃

344 | 345

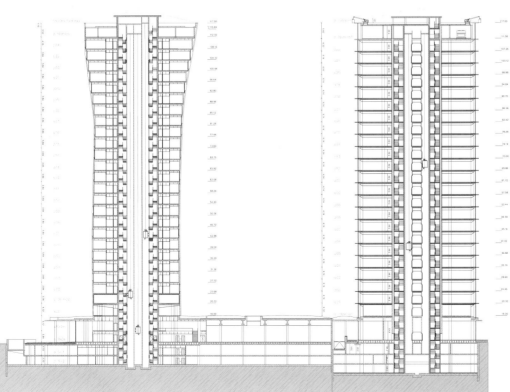

Section
剖面图

Site Plan
总平面图

SHAPE ANALYSIS 造型分析

DIGITAL SHAPE
3D curved surface clad with red aluminum pipe louvers, the hotel volume gradually expands and twists as it reaches the top where the large hotel suites are located.

数字造型
本案呈现出形如塔楼的三维曲面造型。外立面密密麻麻地覆盖着红铝管，集合形成庞大的百页窗外立面，其体量扭曲向上直达顶楼总统套房。

Within the "European Plaza", a pair of twin towers now stands as part of the urban regeneration project for L'Hospitalet, the neighboring city to Barcelona. The site is approximately 8km away from Barcelona El Prat International Airport towards the Barcelona City direction. With this condition of being in between an airport and the urban area, the project constitutes as part of the extension to the "Barcelona Gran Via Fair" Venue, intending to form a gateway that leads people into the fair. Beneath the energetic red twin towers, the entrance hall we designed spreads out to an area of 240,000 m² and is connected to the continuous circulation corridor, the Central Axis.

The project consists of three parts; a hotel tower and an office tower, both about 110 m tall and a lower compartment which has a roof garden that connects the two towers. The hotel tower has a 3D curved surface clad with red aluminum pipe louvers. The volume gradually expands and twists as it reaches the top where the large hotel suites are located. The office tower on the other hand, has a fluid red core in a transparent rectilinear glass box. As the core is set to one side of the tower, the section of this core can be seen externally. Whilst these two towers have clear contrast in form, they acquire a harmonic and complementary relationship.

这座双塔位于"欧洲广场"内，恰好坐落在巴塞罗那和普特拉国际机场之间的位置上，且距离普拉特国际机场仅有 8km 路程，被列入邻近巴塞罗那的 L'Hospitalet 市的城市改造计划中。由于地处巴塞罗那到格兰大道集市延伸线上，本案被认为是一个即将成为通向格兰大道集市的要塞。充满活力的、红艳艳的双塔之下便是被本案设计师扩展出的一块占地 240 000 m² 的大堂。它与连续延伸的走廊——中心轴相连。

本项目由三个部分组成：两座等高 110 m 的酒店塔楼和办公室塔楼，以及与这两座塔楼连接的高度稍低的带楼顶花园的副楼。酒店塔楼被无数红铝管编制而成的百叶窗笼罩在外，呈现出三维曲面造型。整个体量随着塔体一边逐渐开阔变大，一边扭曲向上直达顶楼的总统套房。酒店塔楼旁边的办公室塔楼有一条被呈直线形态的透明玻璃包裹着的图案，呈现水流般的动态美感、渲染上红的娇艳。大部分红色图案集中呈现在办公室塔楼的一侧的玻璃表皮上，其他部分在塔楼其他侧面依稀可见。两座塔楼在外观上，虽然具有鲜明的对比效果，但是两者又呈现出和谐互补的关系。

SKIN ANALYSIS 表皮分析

LIGHT SKIN

Red aluminum tube wrapped around the building, formed the outer epidermis, and has very good adornment effect, reduce the influence of climate on the building.

轻表皮

包裹着整个体量的红色铝管汇聚而成建筑物的外表皮，既起到装饰的效果，又减少了气体对流对建筑的影响。

Ground Floor Plan
首层平面图

2nd Floor Plan
二层平面图

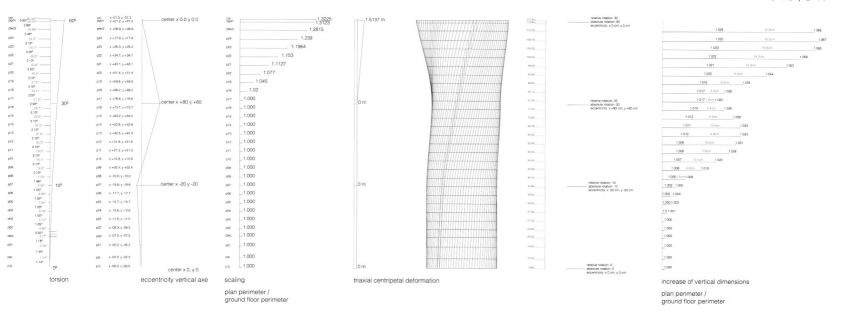

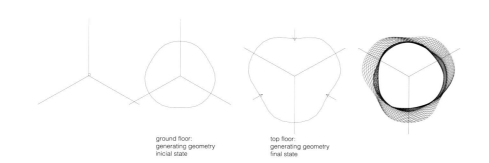

ground floor:
generating geometry
inicial state

top floor:
generating geometry
final state

Ground Floor Parameter
首层参数图

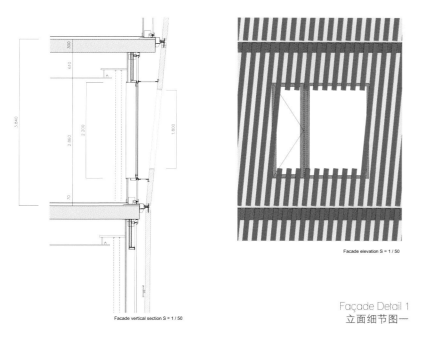

Façade Detail 1
立面细节图一

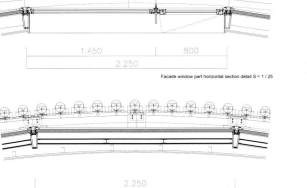

Façade Detail 2
立面细节图二

The hotel tower has the capacity of 345 rooms, to provide for expected visitors of trade fairs as well as for business. The entrance hall, restaurant, and the banquet hall which is also used for conferences or exhibitions are located on the ground floor. On the mezzanine level, there are conference rooms which can be divided into four spaces by sliding partitions and three small meeting rooms. The roof of the lower building volume has a garden terrace that acts as a multifunctional space with a kitchen where cocktail parties can be held. Guests can then migrate to the banquet hall on the ground floor for dinner after the cocktail party, whilst fully appreciating the building as the spaces interlock with each other.

The office tower has 22 office floors, two rentable shops and an entrance hall on the ground floor. The journey from entrance hall to the top level is led by the vertical flows within the red organic form that resembles body organs. The opening sections of elevator halls differ on each floor and from there we can see Mont Juic and the Mediterranean Sea alongside of Barcelona. From an office floor which is able to be divided into four parts for rent, people enjoy the diverse panoramic view of distant streets and mountains of Barcelona.

The city's subway has recently started operating, connecting the major points of the urban regeneration project while the mid-to-high-rise buildings are currently under construction. "Torres Porta Fira" attracts a vast amount of attention as the new landmark is witnessed across Barcelona city and L'Hospitalet.

酒店塔楼有 345 间套房，主要面向商品交易和商业贸易这两大消费群体。首层设置了大堂和餐厅，以及兼具会议和展览功能的宴会厅。夹层设置了多间会议厅，它们具有流动搭配性，可分割成四个小型随意搭配的空间和三个小型会议室。稍矮的副楼楼顶有一个花园露台，是一个多用途空间，还附带方便举办鸡尾酒时使用的厨房。酒会后客人可下楼到首层宴会厅用餐，欣赏这三栋彼此连通的建筑物。

办公室塔楼除首层设置了两个待定出租的商业铺位和大堂以外，其余 22 层均用为办公区域。要从大堂直达顶层，需搭乘形似人体器官的红色电梯垂直上下。因为电梯在不同楼层的开门方向不尽相同，所以你可能有机会一睹邻近巴塞罗那的蒙锥克山的风采，你也可能被映入眼帘的地中海所迷。某一层办公室将被改建成四个部分以便出租，人们可以在此观赏巴塞罗那遥远的城区街道和那连绵起伏的山脉全景。

为了更好地连接城市改造工程中的主要建筑项目和中高层建筑群体，新地铁也与最近启动施工。托雷斯费拉酒店作为新地标而受到广泛关注，同时见证着巴塞罗拉城市和 L'Hospitalet 市的成长。

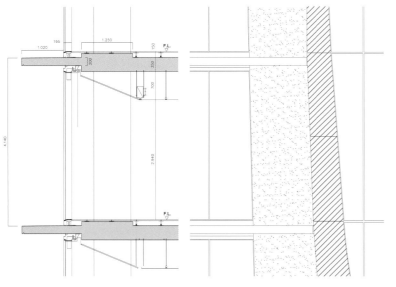

Typical Floor Section
典型楼层剖面图

Façade Elevation
立面细节图

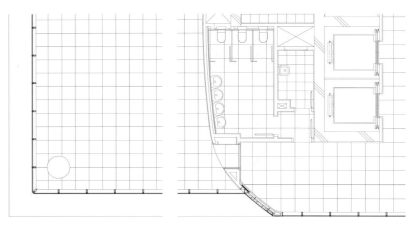

Typical Floor Plan
典型楼层平面图

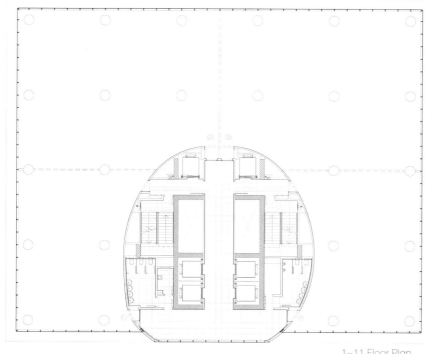

1~11 Floor Plan
1~11 层平面图

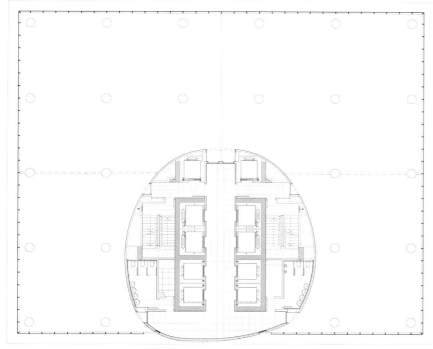

12th Floor Plan
12 层平面图

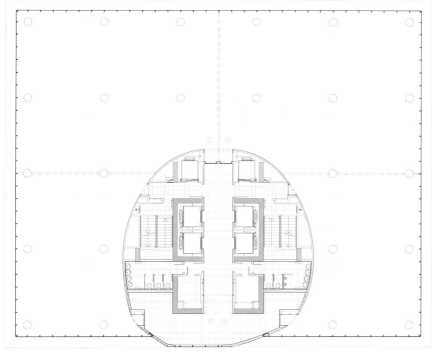

14~22 Floor Plan
14~22 层平面图

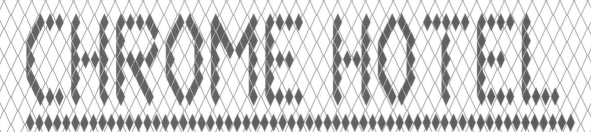

CHROME HOTEL

印度铬金酒店

Architect: Sanjay Puri Architeces
Client: Chocolate Hotels Pvt Ltd
Location: Kolkata, India
Built Area: 3,600 m²

设计公司：Sanjay Puri 建筑事务所
客户：Chocolate Hotels Pvt Ltd
地点：印度加尔各答
建筑面积：3,600 m²

STRUCTURE AND MATERIAL 结构与材料

STRUCTURE
Reinforced Concrete Frame Structure
MATERIAL
Resin Coated Copper Foil, Brick and Glass

结构
钢筋混凝土框架结构
材料
涂胶脂铜箔、砖和玻璃

SHAPE ANALYSIS 造型分析

PICTORAGPHIC SHAPE

Due to space area and height limit, the shape breakthrough lies in outstanding triangle housing structure on the top of the building. Makes the building present a sculpture effect that consists of two cubes tabling in the visual.

象形造型

因场地面积和高度的限制，设计师只能将造型上的突破表现在建筑顶部突出的三角形箱体结构上，使建筑在视觉上形成了一种由两个立方体嵌合而成的雕塑效果。

SKIN ANALYSIS 表皮分析

HEAVY SKIN

The wall of the building is the epidermis part of the building, it is punctuated by small 45 cm diameter circular openingsconcise, and its irregular arrangement have a good adornment effect.

重表皮

建筑的墙体即为建筑的表皮部分，简洁、光滑的墙体上设有直径为 45 cm 的圆形窗口，不规则排列着，形成了很好的装饰效果。

Facing a busy arterial road of the city and flanked by commercial buildings on either side with a residential building at the rear, this small plot for a business hotel had a height limitation of 24 m. The hotel is planned in eight levels with public spaces occupying the first three levels and four levels of rooms above with a rooftop lounge bar on the topmost floor.

The room levels are identified by a rectilinear white block that is punctuated by varying widths of vertical slit windows that cantilevers out over the level of the flyover, forming a wedge at the front corner that houses a suite at each level overlooking a school playground beyond the flyover across the road. And the entire volume comprising of the public spaces and the vertical circulation is punctuated by small 45 cm diameter circular openings.

The hotel is entered through a 24' high lobby with a wall of varied rectilinear composition of wood and glass that curves into the ceiling, slowly fragmenting into individual suspended glass cuboids, creating a sculptural effect.

Suspended within this lobby volume, a wood wrapped corridor acts as an open bar overlooking the lobby while leading into a restaurant at the upper level. A glass punctuated floor with colour change lights echo the exterior wall composition in this open bar corridor with a linear glass bar counter.

Angled trapezoidal planes, punctuated with varied compositions, fold down from the ceiling to create two private dining areas within the restaurant space and fragment the volume into smaller spaces that are lent more privacy. The restaurant design thus creates compositions of form that are varied depending upon which part they are being perceived from.

该酒店面朝繁华的城市主干道，毗邻商业区和住宅区，因此，其开发限高为 24 m。计划将酒店设计为八层，一至三层为公共空间、四至七层为客房，顶层为休闲酒吧。

在建筑外立面一侧，有一个直线型白色箱体结构像窗台一样悬吊在空中。这是一个由于建筑结构恰巧形成的三角形客房，客人站在窗台上可以看到对面学校的操场。同时，为了拓宽视野，整栋建筑物的外墙都设有无数直径为 45 cm 的圆形窗口。

酒店的大堂高约 7 m，以木制直线板和玻璃天花板为装饰，形成了一种独立的悬吊式玻璃立方体，呈现出一种雕刻效果。

悬浮于大堂上空的木质走廊充当着一个独立的开放吧台空间，可以一览大堂和同楼层的餐厅全景。吧台空间的外墙以及其直线型的长吧台与带彩色照明灯光的玻璃地板相互呼应，充满趣味。

带棱角的梯形板从餐厅内的天花板一直延伸至门口，将整个用餐空间分隔成多个小的私密空间，为用餐区笼上了一层神秘的色彩。通过种种精心的设计，整个餐厅在不同的角度呈现出不同姿态，千变万化。

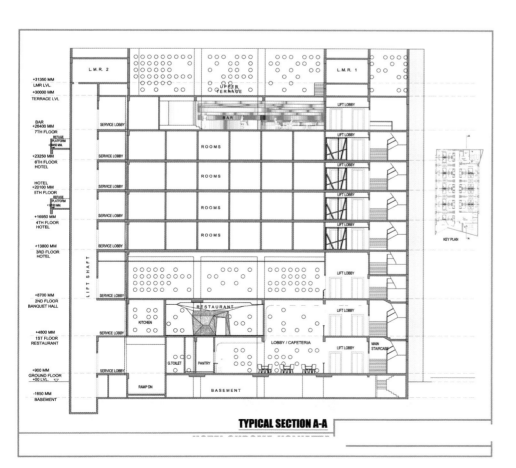

Typical Section A-A
典型剖面图 A-A

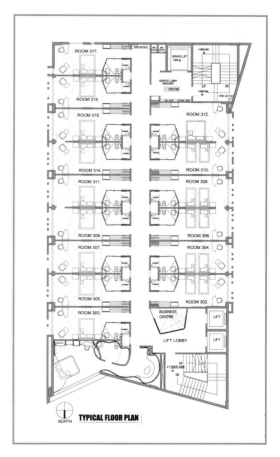

Typical Floor Plan
典型楼层平面图

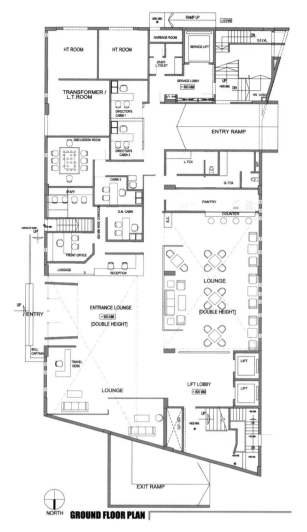

Ground Floor Plan
首层平面图

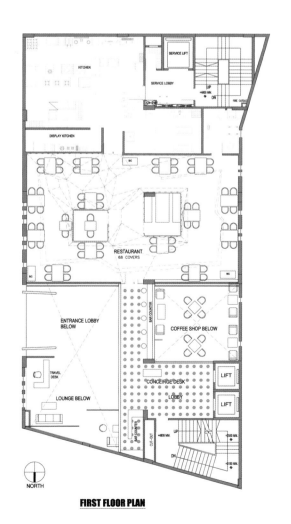

2nd Floor Plan
二层平面图

WILLIAMSBURG HOTEL 威廉斯堡酒店

Architect: Oppenheim Architecture + Design LLP
Location: Brooklyn, New York, U.S.A.
Site Area: 7,989.66 m²
Renderings: Luxigon

设计公司：奥本海姆建筑设计公司
地点：美国纽约布鲁克林区
占地面积：7 989.66 m²
效果图：Luxigon

STRUCTURE AND MATERIAL 结构与材料

STRUCTURE
Framed Structure
MATERIAL
Glass, Steel, Concrete

结构
框架结构
材料
玻璃、钢材、混凝土

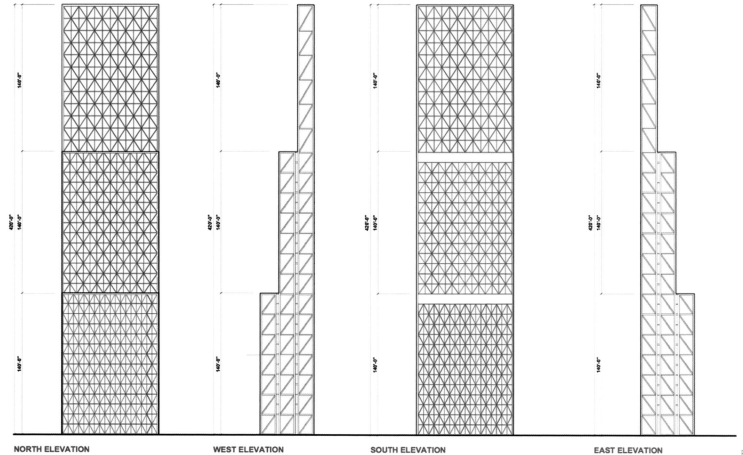

| NORTH ELEVATION | WEST ELEVATION | SOUTH ELEVATION | EAST ELEVATION |

Elevation
立面图

SHAPE ANALYSIS 造型分析

OPEN SHAPE

The use of large area glass allows the hotel guests to have a panoramic view of the Brooklyn and Manhattan skyline.

开放式造型

本案外立面大面积使用透明玻璃，为酒店顾客提供了一个可无障碍观览布鲁克林全景和曼哈顿天际美景的绝佳视野。

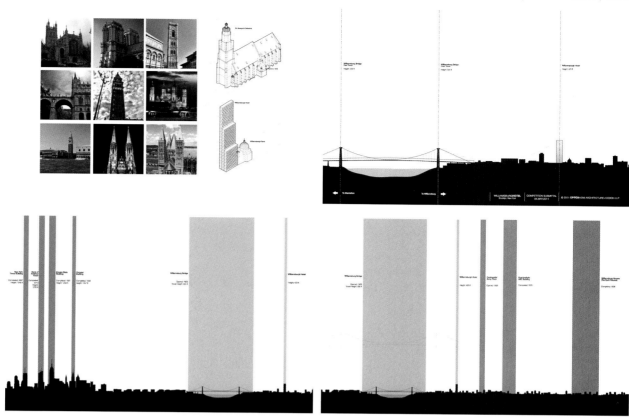

Comparative Analysis
对比分析图

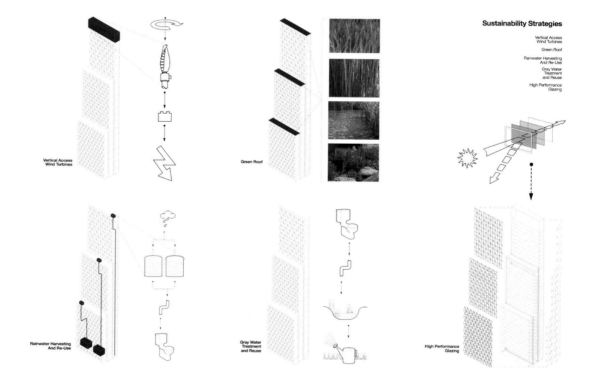

Sustainability Strategies
环保策略

Oppenheim wins international competition to design new hotel in Brooklyn, NY. A third pillar of the Williamsburg Bridge to emerge after 108 years.

Williamsburg, Brooklyn is one of the most interesting and cutting edge neighborhoods in the world. It is a soulful, culinary and style epicenter that is raw, edgy, and visceral. A place that attracts intellectual cognisanti in search of "the real".

The Williamsburg Hotel attempts to capture the essence of this vibrant neighborhood. Adjacent to both the Williamsburg Bridge and the historic Williamsburg Savings Bank, the building, expresses itself as three dramatically proportioned, rectilinear volumes of varied height and materiality. Soaring high above the neighborhood, the hotel becomes the third pillar of the bridge, while serving as an archetypical tower to the domed basilica of the historical bank. With elemental grace and proportion the cluster of towers step back from the street evoking the traditional New York skyscraper typology, but to an extreme scale, with the tallest tower reaching 440' high while maintaining a mere depth of 16'.

The three towers engage and dialogue with the distinctive scales and character of the context, with the lowest volume relating directly to the surrounding neighborhood in both scale and material, the middle one to the adjacent iconic Williamsburg Bank, and the third volume extends to the sky in direct dialogue with the bridge. The slender building becomes a beacon upon entering Brooklyn via car or train, yet scales down towards the street. Like the bridge, the towers are a pure fusion of engineering and architecture, where both internal and external structural systems are expressed and celebrated in a dynamic yet functional pattern. Logical efficiency becomes structural purity, where diagonal steel members serve both to craft an intricately articulated façade, and to optimally resist lateral forces. The faceted envelope allows for incredible views of Manhattan from the hotel rooms, as well as providing an ever changing kaleidoscope that reflects and distorts light. Geothermal, wind, and solar power generation pair with various resource saving strategies to achieve a Platinum LEED rating.

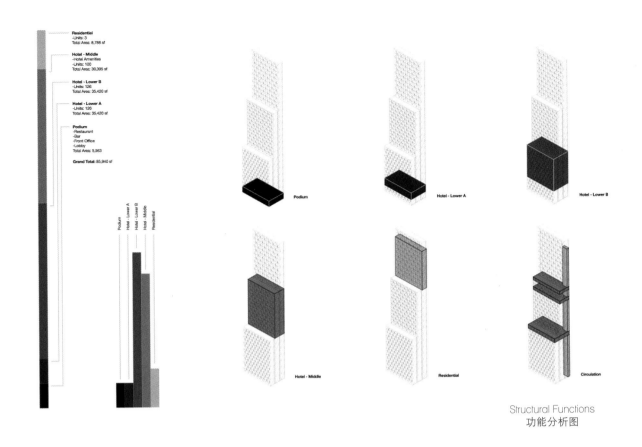

Structural Functions
功能分析图

SKIN ANALYSIS 表皮分析

TRANSPARENT SKIN

The structural system of the building is expressed in the façade design. The biggest highlights of the architectural appearance is that it can reflects the changing light and shadow effect under the different light, while optimize the resist lateral force of the building.

透明表皮

通过晶莹透亮的玻璃外立面，本案精巧的结构系统完美地呈现出来。本案外观设计的最大的亮点在于，它能使人们在不同的光线下，观赏到不同方位的玻璃立面所折射出的不一般的光影效果。此外，玻璃立面还加强了建筑物本身的抗侧向力。

凭借纽约布鲁克林酒店的出色设计，本案设计师赢得了国际设计竞赛的冠军。而本案也成为具有 108 年历史的威廉斯堡大桥的第三大支柱桥梁。

位于布鲁克林的威廉斯堡是世界上最有趣和最尖端的街区之一。这是一个充满人文激情的街区，其美食琳琅满目、风格标新立异、原始之风大行其道、火热激情肆意蔓延、个性率真直抒胸臆，吸引了很多睿智的设计师来此寻找艺术的"真实"。

本案设计尝试捕捉这个街区的活力根源。本案毗邻布鲁克林区第二大标志性建筑威廉斯堡大桥和历史悠久的威廉斯堡储蓄银行，是由三栋高度不同、体量匀称而垂直的塔楼组合而成的建筑。这三栋建筑高耸入云，建筑材质各不相同，但是其体量均按比例递减。酒店成为威廉斯堡大桥的第三大支柱桥墩，同时也成为威廉斯堡储蓄银行圆顶大厅的标志性建筑。优雅的外立面打造了亮眼的气质，成比例的三栋大楼，其体量呈梯级缩进，唤起了人们对纽约传统摩天大厦的记忆。但是，本案极致细腻的结构形态远胜旧日的摩登大厦，最高的塔楼总高虽达 134.1 m，但深度却仅有 4.9 m。

三座塔楼会根据周边环境的风格特色，结合塔楼的独特规模，与街区展开互动和对话。最低的塔楼在规模和材料的选择上，会直接融合周边街区的特色；中高塔楼与标志性建筑威廉斯堡银行相邻，所以规模与选材上务求与之相搭配；最高的塔楼打造了直立上天的形态，与蜿蜒横长的威廉斯堡大桥形成鲜明的对比。最高的塔楼修长耸立，就像一座灯塔，指引布鲁克林大桥上的汽车或火车行驶，并提醒驾驶者在进入街道时应减慢车速。本案不仅仅是一栋纯粹的建筑物，它兼有桥墩的作用，是工程与建筑的创新结合。它的内外结构赋有动态美感，又兼具实用性。建筑的逻辑与功效体现在结构的精纯度上。菱形钢钩件不但能打造出精致的外立面，还能优化建筑物的抗侧向力。立面的玻璃薄膜像个神奇的万花筒一样，或折射或反射出令人惊异的美丽光线，让每一位入住酒店的客人都能透过房间的玻璃薄膜，欣赏曼哈顿的迷人美景。此外，本案还采用了地热、风能、太阳发电能等能节约资源的措施，因此还获得了 LEED（低能电子衍射）白金认证证书。

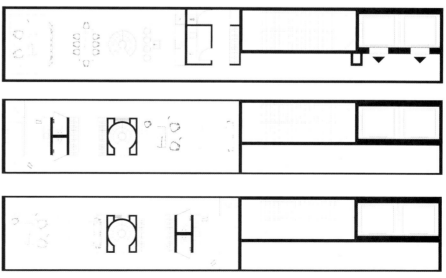

Floor Plan
楼层平面图

SANYA HAITANG BAY MANGROVE HOTEL, HAINAN, CHINA

三亚海棠湾红树林酒店

Architect: ZNA
Client: Antaeus Group
Location: Sanya, China
Built Area: 115,701 m²

设计公司：ZNA 建筑事务所
客户：今典集团
地点：中国三亚
建筑面积：115 701 m²

STRUCTURE AND MATERIAL 结构与材料

STRUCTURE
"Stacked" Framed Structure

MATERIAL
Glass, Concrete, Stone, Wood

结构
"层叠"框架结构

材料
玻璃、混凝土、石材、木材

The site is located in the heart of Shaba Hotel zone in the east corner of the middle Haitang Bay in Hainan: it faces the South Sea and artificial island reef, possessing the excellent view of Wuzhizhou Island; to the east, it is near the fingerlike wetland, and the seafront landscape channel throughout the island provides good availability. There is planned to be open space on the north and south sides, which offers unique confidentiality and exclusive enjoyment, and creates better sea view. The site of this project is intended to be built into a seafront seven star conference resort hotel, with a area of about 21 hectares, and an overall building area of about 115,701 m². The height of architecture is 148 meters, which is obviously higher than that of neighboring construction; therefore, the project shall be unique landmark architecture as to the height of architecture in the region.

本案坐落在海南岛海棠湾中部东隅沙坝酒店区的核心地带：它濒临南海、面朝人造岛礁，享有蜈支洲岛的最佳观景视野；东靠指状湿地，拥有能贯穿全岛的滨海观光隧道，为顾客环岛观光提供便利。本案设计师规划在酒店南北侧打造露天场所，为顾客提供独特的私人空间和专属的奢华享受，同时创造了更好的观海视野。设计师打算将本案打造成一间集度假、会议于一体的七星级滨海酒店。本案占地面积约 210 000 m²，总建筑面积约 115 701 m²。因为它总高 148 m，远远高于周边建筑物，成为当地的地标性建筑。

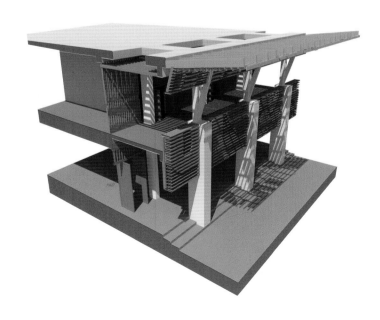

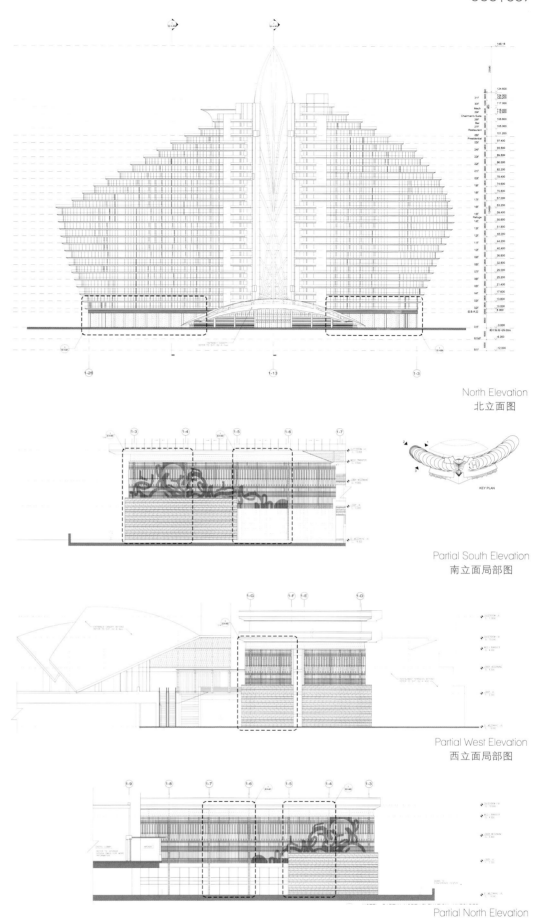

North Elevation
北立面图

Partial South Elevation
南立面局部图

Partial West Elevation
西立面局部图

Partial North Elevation
北立面局部图

SHAPE ANALYSIS 造型分析

PICTOGRAPHIC SHAPE

Shaped like a huge colorful shell. The elegant hotel faces to the demure landscape.

象形造型

本案外形犹如一只巨大的彩色贝壳，优雅娴静地伫立于海边。

SKIN ANALYSIS 表皮分析

LIGHT SKIN

The concept of skin is inspired by local context of Bali. By rearranging the snippets of context, elegant patterns are created with rich enjoyable atmosphere.

轻表皮

酒店表皮图案的设计灵感源自一段巴厘岛的本土文字。本案设计师将文段打乱，重组文字，设计出富有度假风情的典雅图案。

Floor Plan
楼层平面图

摄政埃米尔珍珠饭店

Architect: Shone & Partner Architects, Dennis Lems
Client: Atlas Hospitality, Ahmed S. Al Mutawaa
Location: Abu Dhabi, UAE
Built Area: 124,000 m²
Renderings: Miss 3, Studio Baff

设计公司：Shone & Partner 建筑事务所，Dennis Lems
客户：Atlas Hospitality, Ahmed S. Al Mutawaa
地点：阿联酋阿布扎比市
建筑面积：124 000 m²
效果图：Miss 3、Baff 工作室

STRUCTURE AND MATERIAL 结构与材料

STRUCTURE
Reinforced Concrete Structure

MATERIAL
Steel, Glass, Metal Louvers, Coated Glass Panes

结构
钢筋混凝土结构

材料
钢材、玻璃、金属百叶窗、镀膜玻璃板

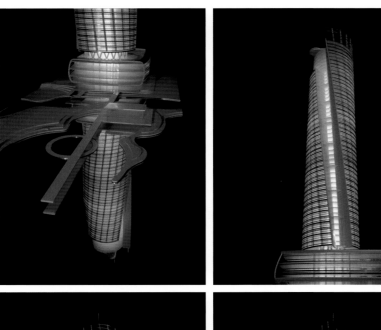

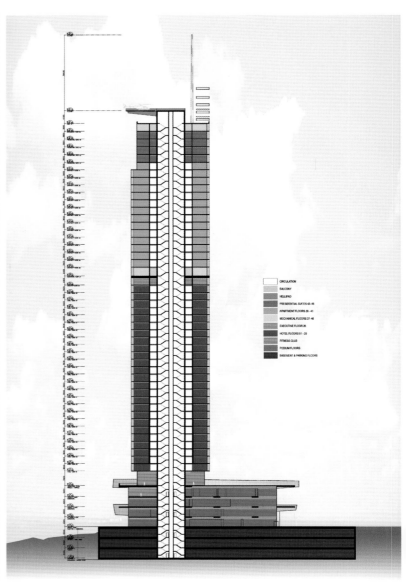

Descriptive Section
细节剖面图

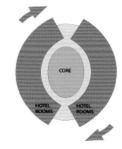
FLOOR 40

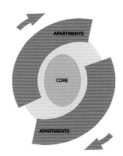
VIEW FROM THE CORRIDORS

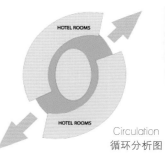
Circulation
循环分析图

SHAPE ANALYSIS 造型分析

DIGITAL SHAPE

Two semi-circular hotel wings spiral upwards around an elliptical circulation core. All the rooms have balconies which define the building's characteristic outward appearance with a linear composition of railing elements.

数字造型

两座半圆形的副楼围绕着一个椭圆形的核心旋转，外观呈现螺旋向上的形态。阳台的设计搭配螺旋式造型，增强了本案雄浑壮阔的气势。由于所有的房间都配有阳台，从整体上看来，阳台扶栏呈现出线性相接的态势。

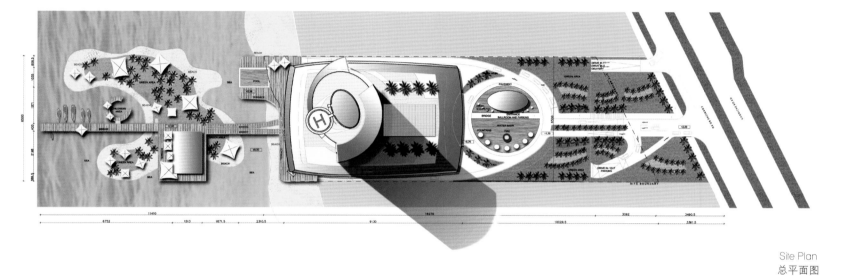

Site Plan
总平面图

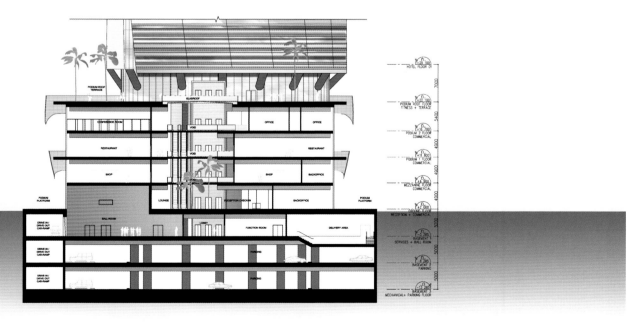

Section
剖面图

In close proximity to the world-famous Emirates Palace, at Corniche Road - the beach promenade - a premium location in the centre of Abu Dhabi, the planned hotel aims to position and prove itself among high-class business hotels.

An expressive form - presenting a unique appearance from every perspective - and a luxurious and modern design concept as well as a comprehensive assortment of amenities, ranging from a helicopter pad with direct elevator access, various restaurants and wellness enticements to a marina complex, create a most comprehensive experience. All in all, Emirates Pearl Hotel has 365 rooms - from double bed to luxury suite - that offer comfortable accommodation for every guest.

本案位于阿布扎比市中心的滨海大道上，紧挨着世界闻名的埃米尔宫。它拥有一线临江美景，享有悠闲美丽的海滨长廊，以跻身世界级顶尖商业饭店之列为目标。

奢华高贵与时尚摩登融为一体的独特外观，给人带来非一般的视觉体验。无论你从什么角度眺望它，入眼之处皆是独一无二的建筑美态。饭店总共设有365间客房，包括标准双人套间和奢华套房，为客人提供最优质的服务与最舒适的居住环境。此外，一应俱全的综合性便利设施，将饭店打造成一个令人难忘的度假胜地，便利的电梯能直达顶楼的直升机停机坪，别具风情的特色餐厅汇聚了各国饮食，配套完善的健身会所提供了舒适的锻炼环境，清爽美丽的海滨小道则让你轻松散步。

SKIN ANALYSIS 表皮分析

LIGHT SKIN

Coated glass panels provide sun protection and open up views, metal louvers complement these elements.

轻表皮

镀膜玻璃板既能反射防晒，还能打造出开阔广大的视野。金属百叶窗与镀膜玻璃板相互配合和补充，能起到增强日光反射的作用，让观光视野更为广阔。

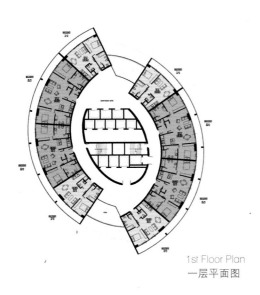

1st Floor Plan
一层平面图

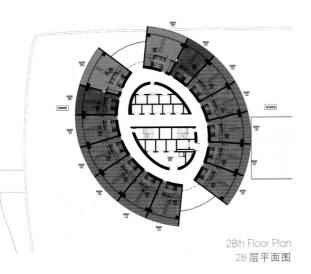

28th Floor Plan
28层平面图

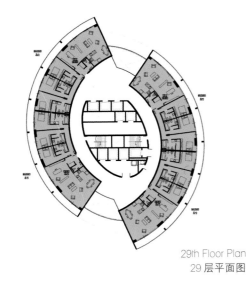

29th Floor Plan
29层平面图

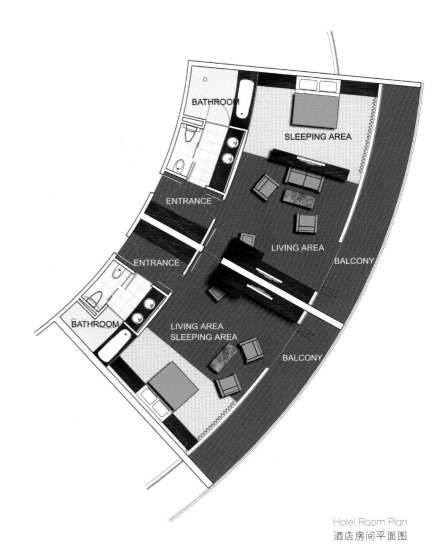

Hotel Room Plan
酒店房间平面图

The ground floor is a completely transparent volume which contains all public and administrative functions of the hotel. Above it rises a five-storey podium volume, which is separated from the towering hotel levels by a second interstice - the Podium Roof - where bar, terrace, pool and jacuzzi are situated.

本案底层采用全透明设计，主要用作公共会客和行政管理。底层之上有个5层楼高的墩座，它与塔形高耸的饭店主建筑不在同一水平线上，它们之间被墩座屋顶撑出了一段间隙，仿佛与饭店主建筑分离了。墩座内设有酒吧、露台、游泳池和按摩浴池等。

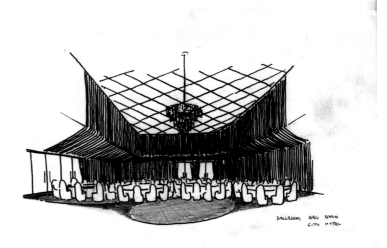

Ballroom
宴会厅

图书在版编目（CIP）数据

100XN建筑造型与表皮. 第2辑. 上 / 先锋空间主编
. — 南京：江苏科学技术出版社，2013.7
ISBN 978-7-5537-1038-9

Ⅰ. ①1… Ⅱ. ①先… Ⅲ. ①建筑－外部－立面造型－世界－图集 Ⅳ. ①TU-881

中国版本图书馆CIP数据核字（2013）第071226号

100XN建筑造型与表皮Ⅱ　上册

主　　　编	先锋空间
责 任 编 辑	刘屹立
特 约 编 辑	陈尚婷
责 任 校 对	郝慧华
责 任 监 制	刘　钧
出 版 发 行	凤凰出版传媒股份有限公司 江苏科学技术出版社
出版社地址	南京市湖南路1号A楼，邮编：210009
出版社网址	http://www.pspress.cn
经　　　销	凤凰出版传媒股份有限公司
印　　　刷	利丰雅高印刷（深圳）有限公司
开　　　本	1 020mm×1 440 mm　1/16
印　　　张	23.5
字　　　数	188 000
版　　　次	2013年7月第1版
印　　　次	2013年7月第1次印刷
标 准 书 号	ISBN 978-7-5537-1038-9
定　　　价	328.00元（USD65.00）（精）

图书如有印装质量问题，可随时向我社出版科调换。